Experimental Glycoscience

Glycochemistry

N. Taniguchi, A. Suzuki, Y. Ito, H. Narimatsu,
T. Kawasaki, S. Hase (Eds.)

Experimental Glycoscience

Glycochemistry

 Springer

Naoyuki Taniguchi, M.D., Ph.D.
Professor, Department of Disease Glycomics, Research Institute for Microbial
Diseases, Osaka University, 1-1 Yamadaoka, Suita, Osaka 565-0871, Japan;
Systems Glycobiology Group, Advanced Science Institute, RIKEN, Wako,
2-1 Hirosawa, Wako, Saitama 351-0198, Japan

Akemi Suzuki, M.D., Ph.D.
Professor, Institute of Glycotechnology, Future Science and Technology Joint Research
Center, Tokai University, Hiratsuka, Kanagawa 259-1292, Japan

Yukishige Ito, Ph.D.
Chief Researcher, RIKEN Discovery Research Institute, Synthetic Cellular Chemistry
Laboratory, 2-1 Hirosawa, Wako, Saitama 351-0198, Japan

Hisashi Narimatsu, M.D., Ph.D.
Research Center for Medical Glycoscience, Advanced Industrial Science and
Technology, AIST Tsukuba, Central 2, Tsukuba, Ibaraki 305-8568, Japan

Toshisuke Kawasaki, Ph.D.
Professor, Research Center for Glycobiotechnology, Ritsumeikan University, 1-1-1
Noji-Higashi, Kusatsu, Shiga 525-8577, Japan

Sumihiro Hase, Ph.D.
Professor, Graduate School of Science, Osaka University, 1-1 Machikaneyama,
Toyonaka, Osaka 560-0043, Japan

ISBN 978-4-431-77923-0 Tokyo Berlin Heidelberg New York
ISBN 978-4-431-77924-7 (eBook)

Library of Congress Control Number: 2008930134

Cover: A ribbon diagram of a legume lectin (*Lathyrus ochrus* isolectin II) complexed with a biantennary glycan (stick model). Image created by Dr. Yoshiki Yamaguchi

Springer is a part of Springer Science+Business Media
springer.com

Typesetting: SNP Best-set Typesetter Ltd., Hong Kong

Printed on acid-free paper

Preface

There is growing interest in glycoscience, which is known to be one of the most important research areas in the medical and life sciences. It is now clear that glycans are implicated in various diseases such as cancer, immune diseases, infectious diseases, muscle-degenerative diseases, neurodegenerative diseases, and diabetes. Glycans also play a pivotal role in fertilization, growth and development, regeneration, and the aging process. Moreover, glycans are likely targets for the discovery of biomarkers, vaccines, and drugs for cancer and inflammatory diseases.

For these reasons, scientists in every field of research have realized that the application of glycoscience contributes to the development of their research projects. They also have a feeling, however, that glycoscience is technically difficult because of the heterogeneity and complexity of glycans. It is evident that many researchers from different disciplines who are not familiar with the specialized techniques used in glycoscience require detailed advice.

Newly developed analytical techniques—namely, the development of mass spectrometry, capillary electrophoresis, and high-performance liquid chromatography—and the accumulated resources of cloned glycosyltransferase genes, lectins, chemically and/or enzymatically synthesized compounds, bioinformatics, and KO mice or gene-targeting animals have opened new avenues for glycoscience.

The Japan Consortium for Glycobiology and Glycotechnology (JCGG) published the Japanese monograph entitled *Glycoscience: A Door to Open the Future* in 2006 written by more than 150 glycoscientists to summarize recent advances in this field in Japan and to offer readers a broad introduction to glycoscience. This present book is the English version but is largely modified and updated. Many of the contributors are experts in their respective fields and have made an effort to present their material in a manner that is understandable to those with general knowledge but with a different background in the biological sciences. The book is presented in three ways: 1) as a "cookbook" with which researchers will be able to prepare the solutions described and follow "recipes" to complete their experiments without needing to consult original research papers; 2) by "concept," which will allow researchers to search out the discipline, research equipment, research device, or methodology required; and 3) as a "table" providing an overview of the research data, among other information. For those who wish to delve into these topics, we included several references at the end of each chapter for further reading. This book will therefore be of interest to many scientists both in glycoscience and in the broader fields of biology, chemistry, and medicine, as well as to postdoctoral fellows, students, and young scientists.

I would like to take this opportunity to introduce the JCGG which was launched 5 years ago (Yoshitaka Nagai, JCGG president, professor emeritus, The University of Tokyo). The organization of this consortium is slightly different from that of the Consortium for Functional Glycomics (CFG) in the United States. Instead of receiving official support from the Japanese government for its operation, scientists took the initiative for creating and supporting the consortium by using their existing government research grants. Scientists who have been funded by different ministries of the Japanese government, such as the Ministry of Education, Culture, Sports, Science and Technology

(MEXT); the Japan Society for the Promotion of Science (JSPS); the Japan Science and Technology Agency (JST); the Ministry of Economy, Trade and Industry (METI); the New Energy and Industrial Technology Development Organization (NEDO); and the Ministry of Health, Labour and Welfare, are involved in this effort. They have joined together to form this consortium and to provide support through their individual research grants.

The JCGG aims to facilitate the exchange of scientific information; the sharing of resources, equipment, and facilities; the fostering of young scientists; and the construction of a database and infrastructures, among other goals. It also aims to launch national research centers in Japan, such as a Systems Glycobiology Center, where glycobiology can merge with such areas as nanotechnology, bioinformatics, and chemical biology to pursue the goals mentioned above, including medical applications. The JCGG holds an annual symposium in which more than 600 people usually participate.

On behalf of the editors, I gratefully acknowledge all those who contributed to this volume despite their busy schedules. Special thanks are due to Mr. Keiichi Yoshida, the secretary general of JCGG, for his skillful assistance in editing. I also thank Kinpodo, Kyoto, who published the original Japanese version, for agreeing to publication of the English version.

Thanks also go to the editorial staff members at Springer Japan, for their constant support, patience regarding the deadline for the manuscript, and skill in directing the production of this book.

I hope this publication will provide an impetus for future research.
On behalf of the editors,

Naoyuki Taniguchi, MD, PhD
February 10, 2008

Editorial Board

Contents

Part 1 Structural Analysis of Sugar Chains

Section I Release of Sugar Chains and Labeling

Section II Sequence Analysis

Contents

Section III Sugar Chain Analysis by Mass Spectrometry

Section IV Analysis of the Three-dimensional Structure of Sugar Chains

Section V Analysis of Sugar–Protein Interactions

Part 2 Chemical Synthesis of Sugar Chains

Section VI Chemical Synthesis of Sugar Chains

Contents of *Experimental Glycoscince: Glycobiology*

List of Contributors

Glossary
Contributed by:
Tadashi Suzuki, Yoshiki Yamaguchi, Shinobu Kitazume

In collaboration with:
Yoko Funakoshi, Kazuyuki Nakajima, Kenji Kanekiyo, Tetsuya Suetake,
Mayumi Kanagawa, Masaki Kato, Kana Matsumoto

Part 1
Structural Analysis of Sugar Chains

Section I
Release of Sugar Chains and Labeling

Chemical Liberation of *N*-linked Oligosaccharides from Glycoproteins

Shin-ichi Nakakita

When *N*-linked oligosaccharides are chemically liberated from the glycoproteins, the hydrazine anhydrous is generally used. Hydrazinolysis has been originally used as a method to determine carboxyl terminal amino acid of protein. Yoshizawa et al. (1966) found that the method is well applicable to chemical liberation of oligosaccharides from glycoproteins. Upon hydrazinolysis, lyophilized samples including glycoproteins are completely solved in hydrazine, and thus, the reaction proceeds smoothly. After hydrazinolysis, released acetyl group of amino sugars, such as GlcNAc and GalNAc, must be subjected to re-acetylation (Takasaki et al. 1982) (Fig. 1).

Hydrazinolysis, differing from an enzymatic procedure to liberate glycans from glycoproteins, is not essentially influenced by chemical structures of glycoproteins, and this makes a basic reason for its reliability in comparison with the enzymatic procedure. Moreover, as described earlier, hydrazine anhydrous shows excellent performance as a solvent. Thus, the method can be applied to not only soluble but also insoluble glycoproteins. Its direct application to animal and plant tissues as well as their cultured cells is also possible, which can be hardly performed by enzymatic methods. A commercial machine for this purpose is available, which is based on a gas-phase reaction. The use of the gas-phase reaction is of practical merit, because it does not require removal of the reagent hydrazine anhydrous. Therefore, *N*-linked oligosaccharides can be liberated from glycoproteins more easily than the corresponding liquid-phase procedure. On the other hand, the gas-phase reaction has difficulty in quantitative liberation of *N*-linked oligosaccharides from heavily insoluble materials like organs and tissues, even with which liquid-phase reaction does work. Once, a commercial machine for liquid-phase hydrazinolysis was marketed. Unfortunately, without it, hydrazinolysis procedure is now exclusively carried out manually. Upon liquid-phase reaction, sealing performance of the screw cap used for this purpose is critical for the success of the reaction. However, it was much better than before, and it seems unnecessary to care about such rigorous decompression sealing for safety operation. However, for the sake of convenience, rapidness and safety, development of an automatic machine for the liquid-phase reaction is necessary.

Hydrazinolysis–*N*-acetylation

1. Sample (glycoprotein or tissue) is lyophilized.
2. Two milligram of powder is put on a bottom of the screw seal tube.
3. Add 0.2 ml of the hydrazine anhydrous in screw seal tube.
4. Mix hydrazine anhydrous and powder.
5. Heat at 100 C, 10 h.

Department of Functional Glycomics, Life Science Research Center, Kagawa Unviersity, Kagawa, Japan

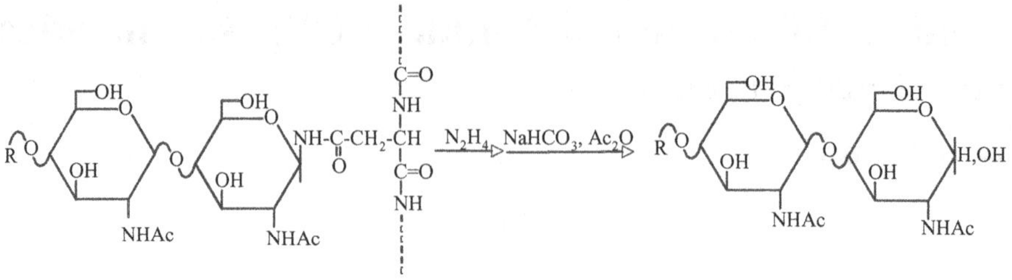

Fig. 1 Chemical liberation of *N*-linked oligosaccharides from glycoproteins by hydrazinolysis-*N*-acetylation

6. Remove hydrazine anhydrous using a rotary pump connected with cold glass trap in vacuo.
7. Add 0.2 mL saturated sodium bicarbonate solution and 0.016 mL acetic anhydride at 0 C for 5 min.
8. Add 0.2 mL saturated sodium bicarbonate solution and 0.016 mL acetic anhydride at 0 C for 30 min.
9. Add to 1 g of Dowex 50 W-X2 (H+ form), and the suspension was poured into a small glass column.
10. The product was washed with 5 column volumes of water.
11. Combine the eluate and the washings, concentrated, and lyophilized. .

References

Takasaki S, Mizuochi T, Kobata A (1982) Hydrazinolysis of asparagine-linked sugar chains to produce free oligosaccharides. In: Ginsburg V (ed) Methods in enzymology, vol 83. Academic Press, New York, pp 263–268

Yoshizawa Z, Sato T, Schmid K (1966) Hydrazinolysis of alpha-1-acid glycoprotein. Biochim Biophys Acta 121:417–420

Release of *N*-glycans by Enzymatic Methods

Noriko Takahashi, Hirokazu Yagi, Koichi Kato

Introduction

To release oligosaccharide moiety from the original glycoprotein molecule, there are two widely used methods: the typical chemical method, hydrazinolysis; and the enzymatic method using glycoamidase A (from almond) or *N*-glycanase F (from *Flabobacterium meningosepticum*). In the past, endo-β-*N*-acetylglucosaminidase used to be applied, too, but the disadavantage of this enzyme is that it always leaves one of the two reducing ends *N*-acetylglucosamine molecules of the oligosaccharide moiety on the substrate protein molecules. On the contrary, glycoamidase A or *N*-glycanase F releases the whole sugar moiety completely from the substrate protein portion. Therefore, the application of the latter is the most convenient method for structural study of oligosaccharide moieties in various glycoproteins.

Glycoamidase A

Glycoamidase A was found in the fruits of sweet almond (Takahashi 1977). It cleaved the acid–amide linkage of the asparagine residue linked to the sugar moiety in the glycoproteins. As a result, the aspartic acid was left in the protein portion, and the same moles of intact oligosaccharide molecules and ammonia were released from the original glycoprotein (Fig. 1). This was a new enzyme, and it was given the new EC number "EC. 3.5.1.52." from the International Enzyme Committee. At that time, similar enzymes could cleave the acid–amide linkage only in small molecules consisting of one Asn linked to one GlcNAc residue. The new enzyme did not have such limitation: it could cleave the acid–amide linkages in arbitrarily big glycoproteins, while preserving the oligosaccharide portions intact. I named the new enzyme "Glycoamidase A", the "A" indicating the source of the enzyme, almond. From that day on, it became widely used for structural study of the oligosaccharide portion of glycoproteins.

We have used the glycoamidase A for about 30 years to determine more than 500 structures of *N*-linked oligosaccharides, as published in over 100 papers.

All *N*-linked oligosaccharides so far tested whether neutral, sialylated, sulfated, or phosphorylated are cleaved from their original glycopeptides by this enzyme. All oligosaccharides recorded on our 2/3D-map database (Takahashi 1977, Takahashi et al. 1995, Tomiya et al. 1988), or our Web application GALAXY is obtained by this glycoamidase A digestion. Now, glycoamidase A is distributed by Seikagaku Kogyo Co., LTD (Tokyo), as "Glycopeptidase A".

Graduate School of Pharmaceutical Sciences, Nagoya City University, 3-1 Tanabe-dori, Mizuho-ku, Nagoya 467-8603, Japan
Phone: +81-52-836-3448, Fax: +81-52-836-3447
E-mail: ntakahas@phar.nagoya-cu.ac.jp

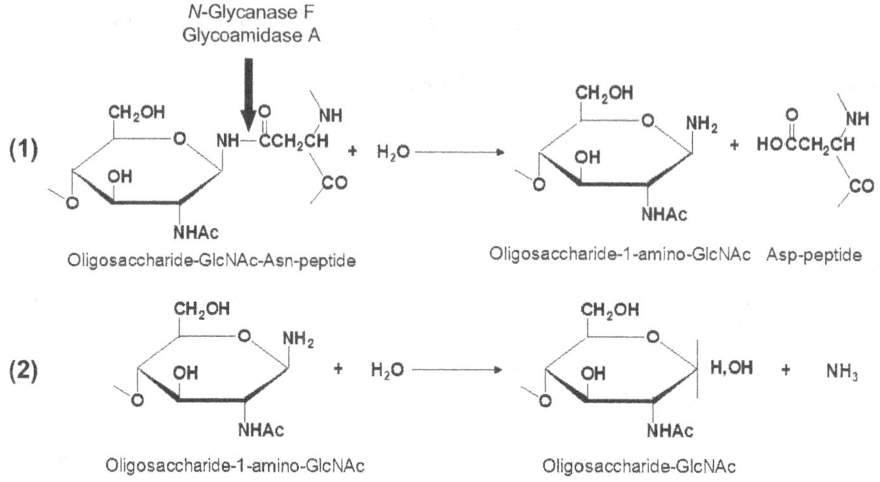

Fig. 1 Reaction mechanism of glycoamidases A and N-glycanase F

Later, 4 years after my discovery of the glycoamidase A from almond, a similar enzyme, N-glycanase F, from *F. meningosepticum* was reported, and now, N-glycanase F is sold by several companies in Japan and the USA under different brand names.

The Reaction Mechanisms of Glycoamidase A and N-glycanase F

The reaction mechanisms of glycoamidases A and N-glycanase F are essentially the same. As shown in Fig. 1, these two enzymes produce three kinds of products (oligosaccharide, peptide, and ammonia). Thus, we can choose any one of these three molecules to assess the enzyme activity.

The Differences of Substrate Specificity Between Two Enzymes, Glycoamidase A and N-glycanase F

The substrate specificity of the two enzymes, A and F, was tried to differentiate using several glycopeptides of animal, plant, and insect origin as substrates.

Glycoamidase A (optimum pH 4) releases all kinds of oligosaccharides from animal, plant, and insect sources, but N-glycanase F (optimum pH 7.5) is applicable only to animal glycoproteins. Structural elucidation of substrates using those two enzymes explained their difference: N-glycanase F could not release oligosaccharides containing α1,3-fucose residue at the asparagine-linked GlcNAc, which are commonly found in plants or insects, including some fruit or grass allergen. Therefore, it is clearly advantageous to use glycoamidase A rather than N-glycanase F for the structural studies of N-glycans of plant or insect origin.

Furthermore, when both enzymes are compared using the same units of each enzyme, it was found that more oligosaccharides, especially those containing sialic acid, were produced in the case of using glycoamidase A. The recovery of oligosaccharides from the digestion of glycoamidase A is usually more than 80% of the original amount.

Fig. 2 How many microheterogenic variations exist in each glycoprotein? One example: Recombinant human soluble Fcγ receptor III (35 kinds of oligosaccharide structures found in the same protein) (Takahashi et al. 2004)

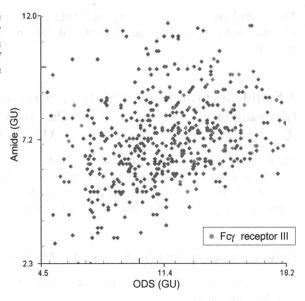

Determination of the Amount of Substrate Protein

If substrate protein amount determination has to be known, then the protein solution should be first dialyzed against water to completely remove inorganic salts or other low molecular weight materials which may be present in the sample. After that, the sample will have to be freeze-dried, but the water molecules tightly bound to the protein must be removed by means of a heated vacuum oven for many days until a constant dry weight is obtained. This is the reason why most people do not determine protein amount this way.

Actually, the absorbance value was determined by measuring the UV-absorption spectrum and weighing the dried sample from it very carefully. Usually 10 mg of a freeze-dried pure protein can be considered to have only 7 mg of the real protein.

The reducing ends of the oligosaccharides are usually derivatized with a fluorescent reagent, 2-aminopyridine, as proposed in Hase's improved method for PA-derivatization of both sialyl and neutral oligosaccharides. Because of its fluorescent nature, the sensitivity of detection of PA-oligosaccharides is at the picomole level.

Contribution of Glycoamidase A to Elucidate the Sugar Microheterogeneity in Glycoproteins

As many data related to oligosaccharide microheterogeneity have been accumulated during the past 30 years, as reported in 2D- and 3D-maps, two typical examples of the oligosaccharide microheterogeneity are shown here.

Example 1. **Recombinant human soluble Fcγ receptor III** (35 kinds of oligosaccharide structures found in the same protein) (Takahashi et al. 2004) (Fig. 2).

Example 2. 40 different sulfated glycans have been successfully mapped by using adequate quantities of sulfated *N*-glycans derived from LS12 cells, which are transfected with sulfotransferase cDNA (Yagi et al. 2005).

These data have shown how many microheterogenic variations exist in each glycoprotein, and without glycoamidase A, it would have been very difficult to make a clear distinction.

A Simple Identification of *N*-glycan Structures Released by Glycoamidase A Digestion, Using Our Web Application GALAXY (Glycoanalysis by the Three Axes of MS and Chromatography)

The database contains approximately 500 kinds of structures of *N*-glycans, all released by glycoamidase A digestion. The structures were elucidated utilizing 2/3D-mapping method and mass spectrometric data. Most of our recently published structures are determined with help of this web application, GALAXY.

Release of *N*-glycans by Enzymatic Methods

Here, we present a procedure for the release of *N*-glycans from commercial human IgG by using Glycoamidase A.

1. Dissolve 15 mg of IgG (corresponding to 200 nmol of oligosaccharide moiety) in 60 µl of distilled water.
2. Heat at 100 °C for 10 min to denature the protein.
3. Adjust the pH of the reaction mixture to 4.
4. Add pepsin (150 µg) and 2 mU of glycoamidase A.
5. Incubate at 37 °C overnight for proteolysis and releasing of oligosaccharide moiety.

At that time and even now, the common belief used to be that glycoamidase A should be given some time (say after overnight) after adding the pepsin. Actually, we confirmed that when pepsin and glycoamidase A were added to the reaction mixture at the same time, there was no decrease of the glycoamidase A activity. Moreover, in this condition (simultaneous addition of glycoamidase A and pepsin, at pH 4, 37 °C, overnight incubating), sialic acid residue was not released from the carbohydrate moiety.

For the glycoamidase A to work well, the size of the peptides must be in the range of 3–40 amino acids. Instead of pepsin (at pH 4), trypsin and chymotrypsin (at pH 8) are also available for the digestion of glycoprotein to such an extent. However, in this case, on **Day 1**, trypsin and chymotrypsin digestion (at pH 8), and on **Day 2**, glycoamidase digestion (at pH 4) should be done.

Day 1. Dissolve 1 mg of IgG in 40 µl of 0.1 M tris–HCl buffer (pH 8.2).

Heat at 100 °C for 10 min to denature the protein.
Then, add trypsin and chymotrypsin (10 µg each), followed by 2 µl of 0.2 M CaCl₂.
Incubate at 37 °C overnight for proteolysis.

Day 2. Heat the reaction mixture at 100 °C for 10 min to inactivate the proteases, followed by evaporation to dryness.

Adjust the pH of the reaction mixture with 1 M citrate–phosphate buffer to pH 4.0 (make sure the pH is 4).
Add glycoamidase A (100–40 µU) and incubate at 37 °C, overnight.

In the special case of some proteins there may be too many disulfide bridge-making proteolysis with trypsin and chymotrypsin digestion difficult. If that were the case, prior treatment of reduction-carboxymethylation would be necessary.

References

Takahashi N (1977) Demonstration of a new amidase acting on glycopeptides. Biochem Biophys Res Commun 76:1194–1201

Takahashi N, Nakagawa H, Fujikawa K, Kawamura Y, Tomiya N (1995) Three-dimensional elution mapping of pyridylaminated N-linked neutral and sialyl oligosaccharides. Anal Biochem 226:139–146

Takahashi N, Cohen-Solal J, Galinha A, Fridman WH, Sautès-Fridman C, Kato K (2002) N-glycosylation profile of recombinant human soluble Fcγ receptor III. Glycobiology 12:507–515

Tomiya N, Awaya J, Kurono M, Endo S, Arata Y, Takahashi N (1988) Analyses of N-linked oligosaccharides using a two-dimensional mapping technique. Anal Biochem 171:73–90

Yagi H, Takahashi N, Yamaguchi Y, Kimura N, Uchimura K, Kannagi R, Kato K (2005) Development of structural analysis of sulfated N-glycans by multidimensional high performance liquid chromatography mapping methods. Glycobiology 15:1051–1060

Release of *O*-glycans by Chemical Methods

Shunji Natsuka

Introduction

Structural research of *O*-glycans has a smaller body of literature compared with that of *N*-glycans, although many physiological functions of *O*-glycans have been reported. This is mainly due to the lack of a suitable technique for structural analysis of trace amount of *O*-glycans.

Conventionally, *O*-glycans are released with alkaline solution through β-elimination which simultaneously reduces the reducing end into a sugar alcohol in order to prevent further degradation. However, the reduced oligosaccharides are not suitable for microanalysis, because of the lack of fluorescence derivatization techniques. One possible answer to this problem is the glycan-releasing reaction of anhydrous hydrazine. Hase and colleagues have optimized hydrazine degradation for *O*-glycan release (Kuraya and Hase 1992), and Iwase and colleagues have optimized gas-phase reaction of hydrazine degradation for *O*-glycan release (Iwase et al. 1992). Following these reports, a group in Oxford also determined reaction conditions for *O*-glycan release (Patel et al. 1993). Since hydrazine degradation can release not only mucin-type glycans but also *N*-glycans and other *O*-glycans, it is a technique adequate for comprehensive analysis of the glycans on glycoproteins (Fig. 1). Semiautomatic equipment for hydrazine degradation and fluorescence-labeling is marketed by HONEN Corporation (Tokyo, Japan). Accumulation of structural data of *O*-glycans from diverse organisms, individuals and organs in various developmental stages, diseases and environments is an important challenge in the near future.

Procedures

1. Dry up a sample completely.
2. Mix the sample with 0.5 ml of anhydrous hydrazine in a vacuum tube.
3. Freeze the sample solution with dry ice–ethanol.
4. Decompress the tube.
5. Heat at 60°C for 6 h.
6. Remove the hydrazine by co-evaporation with toluene.

Department of Chemistry, Graduate School of Science, Osaka University,
1-1 Machikaneyama, Toyonaka, Osaka 560-0043, Japan
Phone: +81-6-6850-5381, Fax: +81-6-6850-5382
E-mail: natsuka@chem.sci.osaka-u.ac.jp

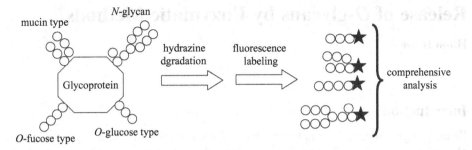

Fig. 1

References

Iwase H, Ishii-Karakasa I, Fujii E, Hotta K, Hiki Y, Kobayashi Y (1992) Analysis of glycoform of O-glycan from human myeloma immunoglobulin A1 by gas–phase hydrazinolysis following pyridyl-amination of oligosaccharides. Anal Biochem 206:202–205

Kuraya N, Hase S (1992) Release of O-linked sugar chains from glycoproteins with anhydrous hydrazine and pyridylamination of the sugar chains with improved reaction conditions. J Biochem 112:122–126

Patel T, Bruce J, Merry A, Bigge C, Wormald M, Jaques A, Parekh R (1993) Use of hydrazine to release in intact and unreduced form both N- and O-linked oligosaccharides from glycoproteins. Biochemistry 32:679–693

Release of *O*-glycans by Enzymatic Methods

Hitoo Iwase

Introduction

Endo-type glycosidase is an indispensable tool for obtaining an intact sugar chain from glycoprotein and also to prepare intact deglycosylated proteins. To release the *O*-glycan sugar chain from mucin-type glycoproteins, endo-α-*N*-acetylgalactosaminidase (endo-GalNAc-ase, glycopeptide α-*N*-acetylgalactosaminidase, EC 3.2.1.97) is usually used. An enzyme from *Streptococcus* (*Dipplococcus*) *pneumonia* is now commercially available as *O*-Glycanase, although its application was fairly restricted due to its narrow substrate specificity. To overcome its narrow substrate specificity, development of a new type of endo-GalNAc-ase having broad substrate specificity was expected. We found two types of endo-GalNAc-ase activities in *Streptomyces* sp. OH-11242; however, we are still awaiting confirmation of the reproducibility of enzyme preparation and its substrate specificity by other researchers (Ishii-Karakasa et al. 1992; Tanaka et al. 1998). Because the secretion of sialyl mucin-type oligosaccharide in normal human urine has been reported, such endo-GalNAc-ase is thought to be present in the natural world (Parkkinen and Finne 1987).

Endo-GalNAc-ases Belong to the Glycoside Hydrolase Family

Endo-GalNAc-ase activity was purified from *S. pneumoniae*, *Clostridium perfringens*, *Alcaligens* sp. and *Bacillus* sp. These enzymes hydrolyzed the glycoprotein having a Core 1 structure (Galβ1,3GalNAc) and released the disaccharide. The specificity was very critical. There are no reports on their cleavage of sugar chains other than the disaccharide (Fig. 1A). The presence of different aglycon specificity between endo-GalNAc-ase from *C. perfringens* and *O*-Glycanase was reported for a compound found in urine from Kanzaki disease (Hirabayashi et al. 1990). It was also well-known that the *O*-Glycanase was inhibited by the presence of EDTA, a high concentration of galactose, chloride ion, and mercury ion. Recently, molecular cloning of the endo-GalNAc-ase from *Bifidobacterium longum* was carried out (Fujita et al. 2005). This indicated the presence of highly conserved open reading frames similar to those of the endo-GalNAc-ases from *C. perfringens* and *S. pneumonia*. Sequence analysis indicated it to be a single peptide having a transmembrane region at both the N- and C-terminal, and a cell wall-binding domain at the C-terminal region. The calculated molecular mass of the enzyme from *Bifidobacterium longum* was 210,269 Da. Because the hypothetical enzyme proteins for *S. pneumonia*, *C. perfringens*, *S coelicolor*, and *E faecalis* by a database search were similar to the enzyme from *B. longum*, they established a novel glycoside hydrolase family 101 including these endo-GalNAc-ases. It is interesting that there is no similarity

Department of Biochemistry, Kitasato University, 1-15-1 Kitasato, Sagamihara, Kanagawa 228-8555, Japan
Phone: +81-42-778-9267, Fax: +81-42-778-8441
E-mail: iwaseh@med.kitasato-u.ac.jp

Fig. 1 Action modes of endo-GalNAc-ase

(A) Conventional Endo-GalNAc-ase activity

|

Natural substrate: Galβ1,3GalNAcα– Ser(or Thr) → Gal β1,3GalNAc + Protein

|

Synthetic substrate: Galβ1,3GalNAcα–PNP → Gal β1,3GalNAc + PNP

(B) Transglycosylation reaction by Endo-GalNAc-ase

Donor: Galβ1,3GalNAcα–PNP or asialo-fetuin

+

Acceptor: Protein, peptide, carbohydrate or other materials with hydroxyl group

↓

Product: Galβ1,3GalNAcα–Acceptor

+

PNP or deglycosylated fetuin

(C) Carbohydrate chain structure of sialyl oligosaccharide released from fetuin by Endo-GalNAc-ase-S treatment

Hexasaccharide: NeuAcα2,3Galβ1,3(NeuAcα2,3Galβ1,4GlcNAcβ1,6)GalNAc

Tetrasaccharide: NeuAcα2,3Galβ1,3(NeuAcα2,6)GalNAc

Trisaccharide: NeuAcα2,3Galβ1,3GalNAc

to any other glycoside hydrolase. Based on the site-directed mutagenesis, two conserved aspartic acids, Asp 682 and Asp 789, were found to be critical for the activity.

Application of Endo-GalNAc-ase

The Core 1 disaccharide, which was also called a Thomsen-Friedenreich (TF) antigen, is abundantly present in mucin-type glycoproteins including mucin. The TF antigen is also well-known as a tumor-associated antigen. Direct demonstration of increased expression of the TF antigen in colonic adenocarcinoma and ulcerative colitis was carried out using *O*-Glycanase (Campbell et al. 1995). After that, this disaccharide was found to be present on a splice variant of CD44v6, which was identified as a tumor metastasis-associated protein (Singh et al. 2006). Quantitation of the TF antigen in the human immunoglobulin A1 subclass was carried out using endo-GalNAc-ase from *Alcaligens* sp. to examine the cause of IgA nephropathy.

Transglycosylation Reaction by Endo-GalNAc-ase

A transfer reaction (reversed hydrolysis) of the Core 1 disaccharide by endo-GalNAc-ase was first found by the transfer of the disaccharide to glycerol, which is added to *O*-Glycanase as a stabilizer (Bardales and Bhavanandan 1989) (Fig. 1B). In addition to the glycerol, Tris, *p*-nitrophenol, threonine, serine, D-glucose, D-galactose, D-fucose and 6-*O*-methylgalactose could become an acceptor. Water insoluble-1-alkanols were also used as the acceptor and produced glycolipids bearing the disaccharide in an α configuration in the co-presence of sodium cholate. In these experiments, *p*-nitrophenyl

Galβ1,3GalNAc (PNP-disaccharide) and asialo-fetuin were used as the disaccharide donor. Unfortunately, the yield of the product remained generally moderate. Therefore, conversion of a glycosidase into a transglycosidase by mutagenesis was expected in the future (Fujita et al. 2005).

Endo-GalNAc-ase-S from *Streptomyces* sp. OH-11242

New type of endo-GalNAc-ase (endo-GalNAc-ase-S) was examined as follows (Iwase and Hotta 1993).

Assay Method of Endo-GalNAc-ase-S

1. Nondegradative mucin (NDM) was prepared from crude porcine gastric mucin by extensive digestion with pepsin.
2. GAGs were removed by CPC precipitation, and NDM was purified by 50–80% ethanol precipitation.
3. Eighty percent ethanol supernatant of the reaction mixture (NDM treated with endo-GalNAc-ase-S) was subjected to TLC.
4. Disaccharide and larger oligosaccharides were separately collected from the TLC plate.

Oligosaccharides were pyridylaminated and their reducing-terminal PA-GalNAc was quantitated. Structures of the sialyl oligosaccharides released from fetuin by endo-GalNAc-ase S are shown in Fig. 1C.

Purification of Endo-GalNAc-ase-S

1. Crude enzyme was prepared as 80% ammonium sulfate precipitation from the culture medium of *Streptomyces* sp. OH-11242.
2. Enzyme was eluted at around isoelectric point 4.9 by gel chromatofocusing.
3. Enzyme from the gel chromatofocusing was eluted at around 0.23 M NaCl by the subsequent DEAE-Toyopearl chromatography.
4. Further purification of the enzyme was carried out by *N*-(*p*-aminophenyl)-oxamic acid-agarose affinity chromatography.

Comment.

If the culture medium of *Streptomyces* sp. OH-11242 contained sufficient amounts of glucose and amino acids as a carbon source, the growth of the bacteria was good, but endo-GalNAc-ase-S could not be well produced (Tanaka et al. 1998).

There is still no report on the preparation of a new-type of endo-GalNAc-ase from other sources. The study of endo-GalNAc-ase-S is still continuing in cooperation with a group at Kyushu University.

References

Bardales RM, Bhavanandan VP (1989) Transglycosylation and transfer reaction activities of endo-alpha-N-acetyl-D-galactosaminidase from Diplococcus (Streptococcus) pneumoniae. J Biol Chem 264:19893–19897

Campbell BJ, Finnie IA, Hounsell EF, Rhodes JM (1995) Direct demonstration of increased expression of Thomsen-Friedenreich (TF) antigen in colonic adenocarcinoma and ulcerative colitis mucin and its concealment in normal mucin. J Clin Invest 95:571–576

Fujita K, Oura F, Nagamine N, Katayama T, Hiratake J, Sakata K, Kumagai H, Yamamoto K (2005) Identification and molecular cloning of a novel glycoside hydrolase family of core 1 type O-glycan-specific endo-alpha-N-acetylgalactosaminidase from Bifidobacterium longum. J Biol Chem 280: 37415–37422

Hirabayashi Y, Matsumoto Y, Matsumoto M, Toida T, Iida N, Matsubara T, Kanzaki T, Yokota M, Ishizuka I (1990) Isolation and characterization of major urinary amino acid O-glycosides and a dipeptide O-glycoside from a new lysosomal storage disorder (Kanzaki disease). Excessive excretion of serine- and threonine-linked glycan in the patient urine. J Biol Chem 265:1693–1701

Ishii-Karakasa I, Iwase H, Hotta K, Tanaka Y, Omura S (1992) Partial purification and characterization of an endo-α-N-acetylgalactosaminidase from the culture medium of *Streptomyces* sp. OH-11242. Biochem J 288:475–482

Iwase H, Hotta K (1993) Release of O-linked glycoprotein glycans by endo-α-N-acetylgalactosaminidase. Methods in Molecular Biology 14:151–159

Parkkinen J, Finne J (1987) Isolation of sialyl oligosaccharides and sialyl oligosaccharide phosphates from bovine colostrum and human urine. Methods Enzymol 138:289–300

Singh R, Subramanian S, Rhodes JM, Campbell BJ (2006) Peanut lectin stimulates proliferation of colon cancer cells by interaction with glycosylated CD44v6 isoforms and consequential activation of c-Met and MAPK: functional implications for disease-associated glycosylation changes. Glycobiology 16:594–601

Tanaka Y, Takahashi Y, Shinose M, Omura S, Ishii-Karakasa I, Iwase H, Hotta K (1998) Screening and fermentation of endo-a-N-acetylgalactosaminidase S, a mucin-hydrolyzing enzyme from *Streptomyces* acting on the GalNAc-O-Ser(Thr) linkage. J Ferment Bioeng 85:381–389

Ceramidase and Related Enzymes

Makoto Ito

Introduction

Ceramide, composed of a fatty acid and a sphingoid base linked by an *N*-acyl linkage, is the common lipid moiety of glycosphingolipids (GSLs). Ceramide is synthesized in the ER, from where it is transported to the Golgi apparatus, and glycosylated in a step-by-step manner by various glycosyltransferases. After recycling from plasma membranes and intracellular organella, GSLs are finally sorted into the lysosomes where they are degraded by various exoglycosidases and ceramidase with the aid of corresponding activator proteins. Ceramidase is an enzyme capable of hydrolyzing the *N*-acyl linkage between a fatty acid and a sphingoid base in a free ceramide. Cermaidase can be classified into three groups (acid, neutral, and alkaline enzymes) depending on their pH optima. It is worth noting that the three types of ceramidase are completely different in primary structure, suggesting that the isotypes of the enzyme are derived from different ancestral genes. This section deals with ceramidases and related enzymes. The action modes of enzymes described in this section are shown in Fig. 1.

Ceramidase

Acid ceramidase—Acid ceramidases showing pH optima at 4.5–5.5 are composed of a 13-kDa α-subunit and a 40-kDa β-subunit. This type of ceramidase is exclusively found in the lysosomes where it participates in the catabolism of ceramide. A genetic dysfunction of the enzyme causes ceramide to accumulate in lysosomes leading to Farber disease, a well-known disorder of sphingolipidosis. Interestingly, knocking out the acid ceramidase gene in mice results in death at the embryonic stage suggesting the enzyme is crucial for the embryonic development of mammals.

Alkaline ceramidase—Alkaline ceramidases showing pH optima at 8.5–9.0 are single polypeptides of relatively small molecular weight. This type of ceramidase is found in the ER of yeasts and humans and seems to participate in regulating the amount of ceramide in the ER, although its physiological functions remain to be uncovered.

Neutral ceramidase—The genetic information for neutral ceramidases showing pH optima at 6.5–8.5 is conserved from bacteria to humans. Interestingly, the neutral ceramidases from vertebrates such as zebrafish, mice, rats, and humans possess a serine/threonine/proline-rich domain located just after the signal/anchor sequence at the N-terminus of the enzyme (Tani et al. 2003). This domain, highly glycosylated with *O*-glycans, is integral to the localization of the enzyme to the plasma membrane as a type II integral membrane protein, i.e., its removal causes the enzyme to detach from the plasma membrane. Furthermore, a hybrid form of GFP with the domain and signal/anchor sequence

Department of Bioscience and Biotechnology, Graduate School, Kyushu University,
6-10-1 Hakozaki, Higashi-ku, Fukuoka 812-8581, Japan
Phone: +81-92-641-2898, Fax: +81-92-641-2907
E-mail: makotoi@agr.kyushu-u.ac.jp

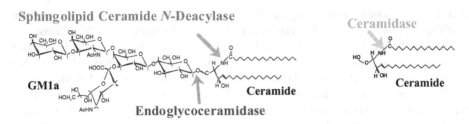

Fig. 1 Action points of endoglycoceramidase, sphingolipid ceramide N-deacylae and ceramidase

Table 1 Classification and characterization of ceramidases

Classification	Optimum pH	Sources	Intracellular localization	Structure and molecular mass	Substrate specificity
Acid ceramidase	4.5	Human, mouse	Lysosomes, secretion	α, β Subunits 50 kDa (13 + 40 kDa)	Cer > dihydroCer
Neutral ceramidase	6.5 ~ 8.5	Bacteria (*P. aeruginosa, M. tuberculosis*), slime mold, fruit fly, zebrafish, mouse, rat, human	Plasma membrane, endosome-like organella, mitochondria, secretion	Single polypeptide 70 ~ 110 kDa	Cer > dihydroCer > phytoCer
Alkaline ceramidase	9.0 ~ 9.5	Yeast (*S. cerevisiae*), human	ER/Golgi	Single polypeptide 30 kDa	PhytoCer > Cer > dihydroCer

Cer ceramide, *dihydroCer* dihydroceramide, *phytoCer* phytoceramide

at the N-terminus remained at the plasma membrane of HEK293 cells. It is worth noting that all neutral ceramidases of bacteria and invertebrates are secreted because they lack the domain at the N-terminus of the enzyme. Neutral ceramidase is involved in the generation of sphingosine and sphingosine 1-phosphate (S1P) on the cell surface and in the extracellular millieum. Interestingly, knockdown of the gene encoding neutral ceramidase affects the formation of heart and blood vessels during the early development of zebrafish, suggesting that the enzyme is essential for S1P signaling (Yoshimura et al. 2004). On the other hand, *Pseudomonas aeruginosa* ceramidase seems to enhance the phospholipase C-induced cytotoxity and to be involved in its pathogenicity (Okino and Ito 2007). Characteristics of the three ceramidases are summarized in Table 1.

Sphingolipid Ceramide N-deacylase (SCDase)

The ceramidases described earlier can hydrolyze the N-acyl linkage in free ceramide but not ceramide bound to glycans (GSLs) or choline phosphate (sphingomyelin, SM). An enzyme that cleaves the N-acyl linkage of ceramides in GSLs as well as in SM was found in bacteria and designated SCDase. The gene encoding SCDase was cloned from *Shewanella alga* G8. Recombinant SCDase degrades various GSLs and SM to generate lyso-forms of sphingolipids (sphingolipids lacking fatty acids) and fatty acids (Furusato et al. 2002). Lyso-forms of sphingolipids are found in cells of patients with some forms of sphingolipidosis and in certain cancer cells; however, homologues of the SCDase gene still have to be identified in mammals. A single SCDase catalyzes a reversible reaction in which the *N*-acyl linkage of the ceramide of GSLs and SM is cleaved or formed under different conditions. The forward (hydrolysis) reaction tends to proceed in the presence of a high concentration of detergents under acidic conditions (0.5~1.0% Triton X-100 at pH 5.0~6.0), whereas the reverse (condensation) reaction tends to proceed in the presence of a low concentration of detergents under neutral condition (0.1% Triton X-100 at pH 7.0). Using these two different reactions, one can produce lyso-forms of various GSLs and SM as well as GSLs and SM with single but different fatty acid molecules. In the latter reaction, if fluorescence- or radioisotope-labeled fatty acids are used, fluorescence- or radioisotope-labeled sphingolipids can be obtained.

Endoglycoceramidase (EGCase)

The EGCase is a unique glycohydrolase that hydrolyzes the *O*-glycosidic linkage between oligosaccharides and ceramides of various GSLs. The prokaryotic EGCase was first found in actinomycetes and then bacteria. Three molecular species of EGCase (EGCases I, II, and III) have been cloned from *Rhodococcus equi*. EGCases I and II hydrolyze ganglio-, lacto/neolacto-, and globo-series GSLs. The hydrolysis of globo-series GSLs by EGCase I was much faster than that by EGCase II. EGCase III (EGALC) specifically hydrolyzes 6-gala series GSLs that are completely resistant to be hydrolyzed by EGCase I and II (Ishibashi et al. 2007). EGCase and a similar enzyme (ceramide glycanase) have been found in not only microorganisms but also various invertebrates. For example, the enzyme has been isolated from the leach, earthworm, short-necked clam, jellyfish, and hydra. However, the enzyme is not likely to be present in mammals. Among these animal EGCases, the genes encoding the enzymes of jellyfish and hydra have been cloned. It is noteworthy that the substrate specificity of animal EGCases reported so far is very similar to that of Rhodococcal EGCase II. In situ hybridization of the EGCase gene in hydra revealed that the enzyme is mainly located in digestive cells of endoderm layers and possibly involved in the digestion of dietary-derived GSLs (Horibata et al. 2004).

Three-dimensional models of EGCases II and III are very similar to the structure of endo-β1,4-glucanase (type 5 cellulase). It is worth noting, however, that EGCases act on GSLs but not on cellulose while endo-β1,4-glucanase acts on cellulose but not on GSLs. The difference in specificity may stem from the shape of the substrate-binding clefts. EGCase seems to be produced from a type 5 cellulase through a change in the shape of the substrate-binding cleft.

Assay for Mammalian Neutral Ceramidase

1. Dissolve 5 μmol of C12-NBD-ceramide (NBD-C12:0, d18:1) in 1 ml of ethanol (5 mM stock solution).
2. Take 10 μl of stock solution in an Eppendorf tube.
3. Add 100 μl of 10% sodium cholate (in MeOH), mix with vortex and dry up under N₂ gas.
4. Add 1 ml of 50 mM Tris–HCl, pH 7.5.
5. Mix with vortex and sonicate for 30 s (substrate solution).

6. Take 10 µl of the substrate solution and transfer to an Eppendorf tube.

7. Add 10 µl of enzyme solution.

8. Incubate at 37°C for 30 min.

9. Heat the sample in a boiling water bath for 5 min to terminate the reaction.

10. Dry the reaction mixture in a Speed Vac concentrator.

11. Re-dissolve in 30 µl of chloroform/methanol (2/1, v/v) and apply an appropriate amount of sample onto a TLC plate (Precoated Silica Gel 60 TLC plate, Merck).

12. Develop the TLC plate with chloroform/methanol/25% ammonia (90/20/0.5, v/v).

13. Analyze and quantify the amount of NBD-dodecanoic acid released and C12-NBD-ceramide remained on the TLC plate with a Shimadzu CS-9300 chromatoscanner (excitation 470 nm, emission 525 nm). Define one enzyme unit as the amount capable of catalyzing the release of 1 µmol of C12-NBD-fatty acid/min from the C12-NBD-ceramide under the conditions described above.

Comment

For bacterial ceramidase, Triton X-100 (final 0.2% in reaction mixture) should be used instead of sodium cholate. Furthermore, calcium chloride should be added to the reaction mixture at the final concentration of 2.5 mM.

References

Furusato M, Sueyoshi N, Mitsutake S, Sakaguchi K, Kita K, Okino N, Ichinose S, Omori A, Ito M (2002) Molecular cloning and characterization of sphingolipid ceramide *N*-deacylase from a marine bacterium, *Shewanella alga* G8. . J Biol Chem 277:17300–17307

Horibata Y, Sakaguchi K, Okino N, Iida H, Inagaki M, Fujisawa T, Hama Y, Ito M (2004) Unique catabolic pathway of glycosphingolipids in a hydrozoan, *Hydra magnipapillata*, involving endoglycoceramidase. J Biol Chem 279:33379–33389

Ishibashi Y, Nakasone T, Kiyohara M, Horibata Y, Sakaguchi K, Hijikata A, Ichinose S, Omori A, Yasui Y, Imamura A, Ishida H, Kiso M, Okino N, Ito M (2007) A novel endoglycoceramidase hydrolyzes oligogalactosylceramides to produce galactooligosaccharides and ceramides. J Biol Chem 282:11386–11396

Okino N, Ito M (2007) Ceramidase enhances phospholipase C-induced hemolysis by *Pseudomonas aeeruginosa*. J Biol Chem 282:6021–6030

Tani M, Iida H, Ito M (2003) O-glycosylation of mucin-like domain retains the neutral ceramidase on the plasma membranes as a type II integral membrane protein. J Biol Chem 278: 10523–10530

Yoshimura Y, Tani M, Okino N, Iida H, Ito M (2004) Molecular cloning and functional analysis of zebrafish neutral ceramidase. J Biol Chem 279:44012–44022

Labeling of Oligosaccharides

Shin-ichi Nakakita

Introduction

It is very difficult to detect a faint amount of oligosaccharide, because generally they have only hydroxyl and aldehyde groups. Thus, many studies have been reported on high-sensitive detection of oligosaccharides by radio-labeling or tagging of a UV-absorbent. In recent years, the method of introducing a fluorescent residue of small molecular weight into the reducing end came to be used worldwide as simplicity and a high-sensitivity detection method of the oligosaccharides. With a small and moderate hydrophobic fluorescent residue, the separability of the labeled oligosaccharides on reversed-phase HPLC remarkably improved. As a result, it is possible that combining several kinds of HPLCs separated up to 500 kinds of oligosaccharides. Such technique is called "Fluorescence labeling of the oligosaccharide". Especially, the method with 2-aminopyridine (Hase et al. 1978) and 2-aminobenzamide (Bigge et al. 1995) as a fluorescent reagent is generally well known. The method of introducing 2-aminopyridine was reported primarily and still used worldwide (Fig. 1).

The oligosaccharide (100 nmol or less) that becomes a sample is dried enough, and the 20 μL of pyridylamino reagent (552 mg 2-aminopyridine dissolved in 200 μL of acetic acid) is added, often mixed, and was heated to 100°C for 1 h. After completion of the coupling reaction, 70 μl of reducing reagent (100 mg borane–dimethylamine complex dissolved in 45 μl of acetic acid and 25 μl of water) is added, often mixed, and was heated to 100°C for 1 h, and the reaction is ended. The excess reagents' removal was used before gel filtration. Recently, removal of excess reagents is used in organic solvent extraction (Yanagida et al, 1999) and subsequent solid phase extraction using a Sep-PAK Plus C18 cartridge (Natsuka et al. 2002), and pyridylamino oligosaccharides can be simply and quickly purified. Recently, the labeling kit for the pyridylamination is sold by Ludger Ltd. (Oxford, UK). Moreover, the automatic pyridylamination machine (GlycoTag™) is sold by Takara Bio Ltd. (Kyoto, Japan). Seventy-six kinds of standard pyridylaminated oligosaccharides are sold by TAKARA BIO INC. and Seikagaku Corporation (Tokyo, Japan). Difference of 2-aminobenzamide hydrophobicity from that of 2-aminopyridine influences the separation on reversed phase HPLC. A 2-aminobenzamide-labeling kit and 22 kinds of the standard labeled oligosaccharides are sold by Ludger Ltd.

Possibly structures of the labeled oligosaccharides are easily determined with comparison of elution positions on reversed phase, size-fractionation, and anion-exchange HPLCs.

Department of Functional Glycomics, Life Science Research Center, Kagawa University, Kagawa, Japan

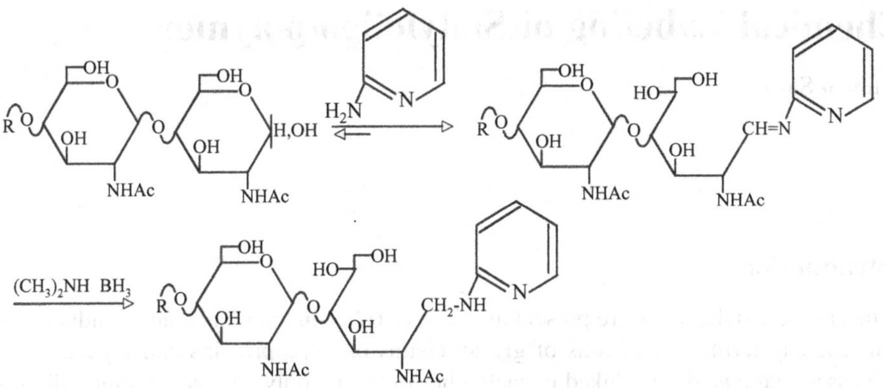

Fig. 1 Labeling of oligosaccharide with 2-aminopyridine by reductive amination

Pyridylamination

1. Infuse saccharide solution in conical microtube.
2. Lyophilize sample.
3. Add 0.02 mL of the coupling reagent solution (552 mg of 2-aminopyridine dissolved in 0.2 mL acetic acid).
4. Mix coupling reagent solution and sample.
5. Heat at 90°C, 1 h.
6. Add 0.07 mL of the reducing reagent solution (200 mg of borane–dimethylamine complex dissolved in 0.08 ml of acetic acid and 0.05 ml of water).
7. Mix reducing reagent solution and sample.
8. Heat at 80°C, 35 min.
9. Mix 0.09 mL water and 0.09 mL phenol/chloroform solution.
10. Centrifuge 2000 × g, 1 min.
11. Remove organic phase.
12. Repeat steps 9–11 twice.
13. Lyophilize water phase.
14. Analyze by reversed-phase HPLC.

References

Bigge JC, Patel TP, Bruce JA et al (1995) Nonselective and efficient fluorescent labeling of glycans using 2-amino benzamide and anthranilic acid. Anal Biochem 230:229–238

Hase S, Ikenaka T, Matsushima Y (1978) Structure analyses of oligosaccharides by tagging of the reducing end sugars with a fluorescent compound. Biochem Biophys Res Commun 85:257–263

Natsuka S, Adachi J, Kawaguchi M et al (2002) Method for purification of fluorescence-labeled oligosaccharides by pyridylamination. Biosci Biotechnol Biochem 66:1174–1175

Yanagida K, Natsuka S, Hase S (1999) A pyridylamination method aimed at automatic oligosaccharide analysis of N-linked sugar chains. Anal Biochem 274:229–234

Chemical Labeling of Sialyloligo/polymer

Chihiro Sato

Introduction

In most cases, sialic acids are present as $\alpha2,3$- or $\alpha2,6$-linked monosialyl residues at the non-reducing terminal positions of glycan chains on glycoproteins and glycolipids. In rare cases, sialic acids are linked to each other to form a polymerized structure, di/oligo/polysialic acid (diSia/oligoSia/polySia). The polySia glycotope exhibits structural diversity in the sialic acid components (Neu5Ac, Neu5Gc, and KDN), in the intersialyl linkages ($\alpha2 \rightarrow 5O_{glycolyl}$, $\alpha2 \rightarrow 8$, $\alpha2 \rightarrow 9$, and $\alpha2 \rightarrow 8/9$) and in the degree of polymerization (DP). Highly sensitive chemical methods to detect di/oligo/polysialic acids were developed (Sato et al. 1998, 1999) with a highly specific reagent for α-keto acid (Hara et al. 1987), and frequent di/oligo/polysialylation on glycoproteins were demonstrated (Sato 2004).

1. Mild Acid Hydrolysis-Fluorometric HPLC Analysis

Mild Acid Hydrolysis and Labeling of Di/Oligo/PolySia Residues with DMB (Fig. 1)

1. Gangliosides, or glycoproteins in solution or blotted on polyvinylidene difluoride (PVDF) membrane (Immobilon P, Millipore, Bedford USA) were hydrolyzed with 0.01 N trifluoroacetic acid at 50°C for 1 h to release di/oligo/polySia unit
2. Freeze-dry the hydrolysates.
3. To the released di/oligo/polySia samples are added 20 µl of 0.01 N trifluoroacetic acid and 20 µl of 7 mM 1,2-diamino-4,5-methylenedioxybenzene (DMB) solution in 5.0 mM trifluoroacetic acid containing 1 M 2-mercaptoethanol and 18 mM sodium hydrosulfite.
4. Incubate at 50°C for 2 h.
5. Apply the resulting supernatants to an HPLC analysis.

Fluorometric HPLC Analysis of DMB Derivatives

1. Set up an HPLC equipped with a Mono or Mini Q HR5/5 (0.5 × 5 cm, GE, Uppsala, Sweden), or Resource Q (1 ml, GE) anion exchange column and a fluorescence detector (FP-2025, JASCO; Tokyo, Japan)
2. Equilibrate the column with 5 mM Tris–HCl (pH 8.0) at 26°C.
3. Apply 2–20 µl of the supernatants to HPLC analysis

Laboratory of Animal Cell Function, Bioscience and Biotechnology Center, Nagoya University, Nagoya 464-8601, Japan
Phone: +81-52-789-4295, Fax: +81-52-789-5228
E-mail: chi@agr.nagoya-u.ac.jp

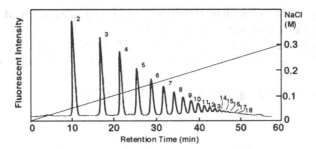

Fig. 1 Mono Q anion exchange chromatography of α2 → 8-linked di/oligo/polyNeu5Ac-DMB. α2 → 8-Linked oligo/polyNeu5Ac was labeled with DMB and applied to a mono Q HR5/5 anion exchange column (0.5 × 5 cm, Cl⁻-form). The column was eluted with 5 mM Tris–HCl (pH 8.0) with a gradient from 0 to 0.3 M NaCl. The elution was monitored by a fluorescence detector (set at wavelength of 373 nm excitation and 448 nm emission). Each peak is assigned from the order of elution, based on those of Neu5Ac-DMB and Neu5Acα2 → 8Neu5Ac-DMB. (Neu5Ac)₂-DMB, (Neu5Ac)₃-DMB, ..., (Neu5Ac)₇-DMB represent DMB derivatives of α2 → 8-linked Neu5Ac dimer, trimer, ..., heptamer, respectively

4. Elute the labeled di/oligo/polySia at a flow rate of 1.0 ml/min with a linear gradient of NaCl (0–60 min, 0 → 0.3 M) after 20 min wash with 5 mM Tris–HCl (pH 8.0) (flow-through).
5. Detect the fluorescence of the elute with a fluorescence detector at excitation 373 nm and emission 448 nm (Fig. 1).

Comments

Using this method, we can detect up to 20 DP. Carbopac-PA1 enables us to determine higher oligomers if you require.

Ammonium acetate can be used instead of NaCl.

2. Procedures for the Fluorometric C_7/C_9 Analysis

Periodate Oxidation/Reduction/Hydrolysis/DMB-Labeling

1. Prepare the reagents. Solutions A to G are prepared as follows: Solution A, 40 mM sodium acetate buffer (pH 5.5); Solution B, 0.25 M periodate; Solution C, 3% ethylene-glycol; Solution D, 0.2 M sodium borohydride in 0.2 M sodium borate buffer (pH 8.0); Solution E, 0.2 M trifluoroacetic acid; Solution F, 0.01 M trifluoroacetic acid; Solution G, 7 mM DMB in 5 mM trifluoroacetic acid containing 1 M 2-mercaptoethanol and 18 mM sodium hydrosulfite. Solutions B and D were freshly prepared. Solutions A, D, E, and F are stored at 4°C. Solution G can be used for 2 weeks when stored at 4°C.
2. For glycoproteins, glycolipids or oligosaccharides in solution, samples (0.25–1,000 ng as Sia) are dissolved in 25 μl of Solution A, and 2 μl of Solution B are added.
3. After incubating at 0°C for 3 h in the dark, 5 μl of Solution C and 32 μl of solution D are added successively and allowed to stand at 0°C overnight.
4. To the resultant mixture is added 1 μg of pyruvic acid as an internal standard for quantitating the resultant sialic acids and set volume to 100 μl with water.

5. Following further addition of 100 μl of Solution E to adjust to 0.1 M trifluoroacetic acid, the mixture is hydrolyzed at 80°C for 1 h.

6. The hydrolysate is dried up by a Speed Vac (SAVANT Instrument, NY, USA).

7. After addition of 20 μl of the solution F, 20 μl of solution G is added to the samples and incubated at 50°C for 2 h.

8. Apply the resulting supernatants to an HPLC analysis.

Comments

For glycoproteins on the PVDF membrane, the glycoprotein band is excised from the membrane, and cut into small pieces. The membrane pieces were soaked in 250 μl of Solution A and 20 μl of Solution B is added. After keeping at 0°C for 3 h in the dark, 50 μl of Solution C and 320 μl of solution D are added successively and incubated at 0°C overnight. After discarding the solution, the membrane pieces are washed three times with 1 ml of water each. The resultant membrane pieces are hydrolyzed, together with 1 μg of pyruvic acid added as an internal standard, in 200 μl of a half-diluted Solution E at 80°C for 1 h. The membrane pieces are removed, and the solution was dried up by the Speed Vac.

Fluorometric HPLC Analysis

1. Set up an HPLC equipped with a TSK-gel ODS-120T column (250 × 4.6 mm *i.d.*, Tosoh), and a fluorescence detector (FP-2025, JASCO).
2. Equilibrate the column using methanol/acetonitrile/water (7:9:84, v/v/v) at 26°C.
3. 2–20 μl of the supernatants to HPLC analysis.
4. Elute the labeled sialic acid isocratically at a flow rate of 1.0 ml/min.
5. Detect the fluorescence of the elute with a fluorescence detector at excitation 373 nm and emission 448 nm (Fig. 2).
6. Estimate the ratios of the quantity of internal sialic acid residues (C_9(Sia)) to that of total sialic acid residues (C_7(Sia)+C_9(Sia)).

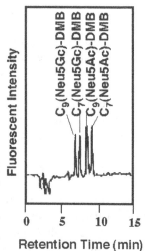

Fig. 2 A typical elution profile of DMB derivatives of C_7-analogues and authentic sialic acids (C_9) on the fluorometric HPLC. Disialyl silalitols, 12.5 ng each of Neu5Acα2 → 8Neu5Acα2 → 8-Neu5Ac-ol and Neu5Gcα2 → 8Neu5Gcα2 → 8-Neu5Gc-ol, were subjected to the periodate oxidation/reduction/hydrolysis, DMB derivatization and fluorometric HPLC on a TSK-gel ODS-120T column (250 × 4.6 mm² *i.d.*). The column was eluted with methanol/acetonitrile/water (7:9:84, v/v/v) at 1.0 ml/min at 26°C. Elution profile was monitored by measurement of fluorescence: excitation, 373 nm; emission, 448 nm

Comments

The following limitations are noted in the application of this method. (1) This method is applicable only to $\alpha2 \rightarrow 8$-linked oligo/polymer of N-acylneuramininc acid and cannot be used for DP analyses of $\alpha2 \rightarrow 9$, $\alpha2 \rightarrow 8/\alpha2 \rightarrow 9$-mixed linkage polymers or $\alpha2 \rightarrow 5O_{glycolyl}$-linkage. (2) The C_9-derivatives formed do not always arise from $\alpha2 \rightarrow 8$-linkages, because 8-O-substituted Neu5Acyl residues may also give the same C_9-derivatives. Therefore, mild alkali-treatment of samples is usually carried out prior to periodate oxidation. (3) The molar proportion of C_9-derivatives to C_7-derivatives does not directly represent the DP unless it is a linear polySia chain. Thus, the method does not yield the DP for multiple sialylated chains present in the same sample. In general, glycoproteins have more than one glycan chains, and even one glycan chain bears 2–4 non-reducing terminal residues that may be terminated by monoSia or oligoSia residues.

References

Hara S, Takemori Y, Yamaguchi M, Nakamura M, Ohkura Y (1987) Fluorometric high-performance liquid chromatography of N-acetyl- and N-glycolylneuraminic acids and its application to their microdetermination in human and animal sera, glycoproteins, and glycolipids. Anal Biochem 164:138–145

Sato C (2004) Chain length diversity of sialic acids and its biological significance. Trends Glycosci Glycotech 16:331–344

Sato C, Inoue S, Matsuda T, Kitajima K (1998) Development of a highly sensitive chemical method for detecting $\alpha2 \rightarrow 8$-linked oligo/polysialic acid residues in glycoproteins blotted on the membrane. Anal Biochem 261:191–197

Sato C, Inoue S, Matsuda T, Kitajima K (1999) Fluorescent-assisted detection of oligosialyl units in glycoconjugates. Anal Biochem 266:102–109

Separation of Oligosaccharides by 2D HPLC

Shunji Natsuka

Introduction

A method for separation and identification of fluorescence-tagged oligosaccharides by two kinds of chromatography was reported by Hase in 1978 (Hase et al. 1978). Since then the microanalysis of labeled glycans has been greatly improved through the use of HPLC, and a sensitivity of glycan detection at the femtomole level (10^{-15} mol) has been obtained.

As shown in Fig. 1, this method, which is called two-dimensional (2D) HPLC-mapping, is a technique for structure analysis through the comparison of elution positions on two kinds of HPLC column with those of standard glycans. This method is used by many researchers worldwide and contributes to the structural analysis of various glycans. In particular, the 2D HPLC-mapping method is an excellent method for isomer distinction in trace amount of glycans. Furthermore, even a beginner can perform this analysis relatively easily, because many standard pyridyamino glycans are marketed by the Seikagaku Corporation and Takara Bio Inc. If standard pyridylamino glycans of the mucin-type come to the market, it is expected that the structural study of mucin-type glycans will also markedly progress. Recently, a 2D HPLC-mapping method for glycosaminoglycans was reported by Takagaki and others (Iwafune et al. 2004). This method used indexes of molecular size and degree of sulfation.

Glucose units are used as a numerical index of elution positions on reversed phase HPLC in many cases in the current 2D HPLC-mapping method. However, this value fluctuates upon small changes in conditions of solvent, temperature, column age, etc. In contrast, a conversion index called the reversed phase scale which has been devised by Yanagida et al. (1998) make it possible to compare the elution position on reversed phase HPLC with high accuracy. In the future, it is expected that the structural study of glycans will widely spread with the advent of HPLC apparatus specialized for 2D HPLC-mapping, the kit of standard glycans for the reversed phase scale, and further expansion of the reversed phase scale database.

Procedures

1. Inject a sample (e.g., PA–glycan mixture) into a TSKgel Amide-80 column (0.46×7.5 cm^2, Tosoh) with the solvent (80% acetonitrile and 20% 50 mM ammonium formate, pH 4.4) at 0.5 ml/min of a flow rate, and at 25°C of a column temperature.
2. Reduce the acetonitrile content to 65% in 5 min, 55% in another 5 min, and then 30% in the next 20 min.
3. Fractionate the separated samples with time or peaks.
4. Concentrate the fractions.

Department of Chemistry, Graduate School of Science, Osaka University,
1-1 Machikaneyama, Toyonaka, Osaka 560-0043, Japan
Phone: +81-6-6850-5381, Fax: +81-6-6850-5382
E-mail: natsuka@chem.sci.osaka-u.ac.jp

Fig. 1 Comparing map positions of standard and sample glycans

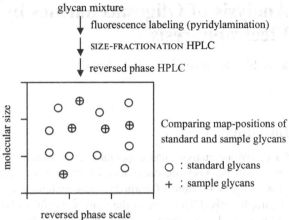

5. Inject the fractions into a Cosmosil 5C18-P column (0.2×25 cm^2, Nakarai Tesque) with the solvent (0.075% 1-butanol in 50 mM triethylamine acetate, pH 4.0) at 0.2 ml/min of a flow rate, and at 25°C of a column temperature.
6. Increase the 1-butanol content to 0.5% in 105 min.
7. Convert the retention-times of the peaks by appropriate index (e.g., the reversed phase scale referred in 3)
8. Plot the 2D HPLC map.

References

Hase S, Ikenaka T, Matsushima Y (1978) Structure analyses of oligosaccharides by tagging of the reducing end sugars with a fluorescent compound. Biochem Biophys Res Commun 85:257–263
Iwafune M, Kakizaki I, Nakazawa H, Nakatsuka I, Endo M, Takagaki K (2004) A glycomic approach to proteoglycan with a two–dimensional polysaccharide chain map. Anal Biochem 325:35–40
Yanagida K, Ogawa H, Omichi K, Hase S (1998) Introduction of a new scale into reversed–phase- high–performance liquid chromatography of pyridylamino sugar chains for structural assignment. J Chromatogr A 800:187–198

Analysis of Oligosaccharides by Capillary Electrophoresis

Mitsuhiro Kinoshita[1], Kazuaki Kakehi[2]

Introduction

Carbohydrate analysis has a special feature which should be capable of high-resolution separation, because oligosaccharides derived from glycoconjugates are usually composed of a complex mixture of carbohydrates including isomers and homologues. Capillary electrophoresis (CE) is one of the most powerful techniques in terms of resolving power and is applied to the analysis of various carbohydrates from glycoconjugates. A combination of CE and laser-induced fluorescence (LIF) detection enables to detect even fmol (10^{-15} mol) to amol (10^{-18} mol) of carbohydrates. Multiplexed CE systems and multi-channel microfluidic devices have been developed for high-throughput research and are applied to clinical analysis of serum glycome.

Principle

Capillary electrophoresis separation is based on the migration of analytes in fused-silica capillaries filled with an electrolyte. Separation is usually performed using a capillary of 50–100 μm internal diameter, under an electric field of several hundred voltage per centimeter capillary. Under such high voltage, carbohydrates are migrated on the basis of their charge to mass ratios, and directly detected by an on-column detector. Although sample diffusion in liquid chromatography often hampers resolution, CE using a very narrow capillary can overcome such problems due to low-sample diffusion in the electrolyte, and achieve high-resolution separation. CE is available in various separation modes such as simple zone electrophoresis, micellar electrokinetic chromatography, and gel electrophoresis. The different modes of CE are easily achieved simply by changing the electrolyte and/or the capillary column. These features of CE can extend its applicability to various kinds of carbohydrates including monosaccharides, free or glycoprotein-derived oligosaccharides, and glycosaminoglycans (Kakehi and Honda 1996; Kakehi et al., 2002). Here, we show some typical examples for the analyses of hyaluronan (HA)-derived oligosaccharides and IgG-derived oligosaccharides.

Faculty of Pharmaceutical Sciences, Kinki University, Kowakae 3-4-1, Higashi, Osaka 577-8502, Japan
Phone: +81-6-6721-2332, Fax: +81-6-6721-2353
E-mail: [1]m-kino@phar.kindai.ac.jp, [2]k_kakehi@phar.kindai.ac.jp

Procedures

Preparation of HA Oligosaccharides

1. Hyaluronan (100 mg) from *Streptococcus zooepidemicus* was dissolved in 0.15 M citrate/HCl (pH 5.3, 10 ml), digested at 37°C overnight with ovine testicular hyaluronidase (2,000 units), and lyophilized to dryness.
2. The lyophilized material (1 mg) was dissolved in distilled water (1 ml), and the solution was filtered through an ultrafiltration membrane (10 kDa cut-off; Millipore) and a portion (10 µl) was analyzed on a Beckman CE apparatus equipped with a UV detector.

Preparation of 2-AA Labeled Carbohydrates

1. To a sample of oligosaccharide (ca. 0.1–1 nmol) or a mixture of oligosaccharides (typically obtained from 10–1,000 µg of glycoprotein samples), is added a solution (200 µl) of 2-aminobenzoic acid (2-AA) and NaBH$_3$CN, prepared by dissolution of both reagents (30 mg each) in methanol (1 mL) containing 4% CH$_3$COONa and 2% boric acid.
2. The mixture is kept at 80°C for 60 min. After cooling, water (200 µl) is added to the reaction mixture.
3. The reaction mixture is applied to a small column (1 cm × 30 cm) of Sephadex LH-20 equilibrated with 50% aqueous methanol. The earlier eluted fractions which contain labeled oligosaccharides are collected and evaporated to dryness. The residue is dissolved in water (100 µl), and a portion (typically 10 µl) is used for CE.

Capillary Electrophoresis for HA-derived Glycans

1. CE is performed with a capillary electrophoresis system (Beckman P/ACE 5010) equipped with a UV detector (a 200 nm-filter). Separation is performed using an inner surface modified capillary (typically, a DB-1 capillary available for gas chromatography with 100 µm i.d., 27 cm effective length).
2. Prior to the analysis, the capillary is rinsed with 50 mM Tris–borate buffer (pH 8.3) for 1 min at 20 p.s.i.
3. The capillary is filled with the same buffer containing 10% polyethylene glycol (average MW 70,000, PEG70000) for 1 min at 20 p.s.i.
4. A sample solution containing HA oligosaccharides is introduced for 10 s at 1.0 p.s.i.
5. Analysis is performed under a constant voltage mode of 6 kV.

Capillary Electrophoresis for Fluorescent-Labeled IgG-derived Glycans

1. CE is performed with a capillary electrophoresis system (Beckman P/ACE MDQ Glycoprotein system) equipped with a He–Cd laser-induced fluorescence detector. Separation is performed using an inner surface modified capillary (typically, a DB-1 capillary available for gas chromatography with 100 µm i.d., 20 cm effective length).
2. Prior to the analysis, the capillary is rinsed with 50 mM Tris–borate buffer (pH 8.3) for 1 min at 20 p.s.i.

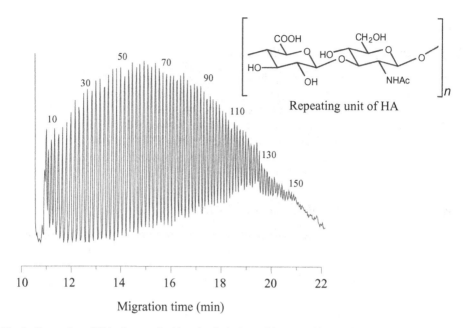

Fig. 1 Separation of HA oligosaccharides. Analytical conditions: capillary, DB-1 capillary (27 cm total length, 20 cm effective length, 100 μm i.d.); running buffer, 50 mM Tris–borate (pH 8.3) containing 10% PEG70000; applied voltage, 6 kV (reverse polarity); sample injection, pressure method (0.5 p.s.i., 10 s). Detection, UV absorption at 200 nm

3. The capillary is filled with the same buffer containing 10% PEG70000 for 1 min at 20 p.s.i.

4. A sample solution containing 2AA-labeled IgG-derived glycans is introduced for 10 s at 0.5 p.s.i.

5. Analysis is performed under constant voltage mode of 25 kV (reverse polarity).

6. Fluorescence detection is performed with a 405 nm emission filter by irradiating with a He–Cd laser-induced 325 nm ultraviolet light.

Determination of Molecular Masses of Acidic Polysaccharides

Glycosaminoglycans (GAGs) are linear polysaccharides composed of repeating disaccharide units of uronic acid and N-acetylhexosamine, and are highly negative-charged macromolecules. Their molecular masses are varied with physiological conditions and are often required to obtain the accurate molecular masses or distribution of the molecular species for understanding their biological functions.

The CE using an electrolyte containing polyethyleneglycol as dynamic sieving material allows resolution of GAG oligomers/polymers based on their molecular masses with ultra high resolution. An example of the separation of hyaluronic acid (hyaluronan, HA) is shown in Fig. 1.

When Tris–borate buffer (50 mM, pH 8.3) containing 10% polyethyleneglycol 70000 is used as the running buffer, HA molecular species composed of more than 150 disaccharide units can be discriminated within 25 min as shown in Fig. 1. To minimize electroosmotic flow (EOF) for high-resolution separation, a capillary (DB-1) of which inner

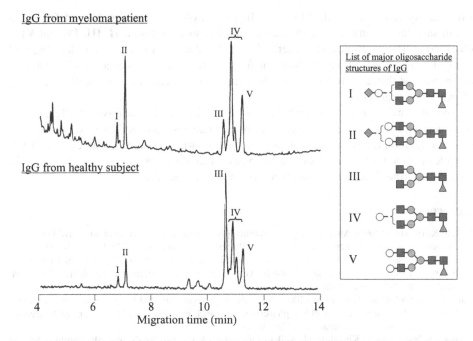

Fig. 2 Analysis of N-linked oligosaccharides in IgG. Analytical conditions: capillary, DB-1 capillary (30 cm total length, 20 cm effective length, 100 μm i.d.); running buffer, 50 mM Tris–borate (pH 8.3) containing 10% PEG70000; applied voltage, 25 kV (reverse polarity); sample injection, pressure method (0.5 p.s.i., 10 s). Fluorescence detection was performed with a 405 nm emission filter by irradiating with a He–Cd laser-induced 325 nm light. The structures of major oligosaccharide in IgG are shown at *right panel*. *Left upper panel* is the profile of oligosaccharides in IgG from myeloma patient, and *left lower panel* is those of healthy subject. The symbols were: GlcNAc (*filled square*), Man (*filled circle*), Gal (*open circle*), Fuc (*filled triangle*), and NeuAc (*open rhombus*)

surface is chemically modified with dimethylpolysiloxane should be used (Kakehi et al. 1999).

Analysis of Oligosaccharides in Glycoproteins

Glycosylation is one of the most important post-translational modifications of proteins, and involved in expression of biological functions such as signaling, protein folding, and molecular interactions. Therefore, we have to analyze carbohydrate-modification of proteins for understanding of protein functions in post-genome era.

The CE with laser-induced fluorescence detection (LIF) is a powerful and useful tool for ultra-high sensitive detection of oligosaccharides. An example of the analysis of oligosaccharides in IgG heavy chain is shown in Fig. 2.

IgG contains the agalacto-biantennary glycan (**III**) as the major glycan, and **IV** and **V** with one Gal residue to one of the GlcNAc residues. Asialo-biantennary glycan (**V**) is also observed. A small amount of mono sialo-biantennary glycans (**I** and **II**) is also observed at the earlier migration times (ca. 7 min). In Fig. 2, the IgG bands resolved by SDS PAGE is "in-gel" digested with *N*-glycoamidase F, and the released *N*-glycans are analyzed after fluorescent derivatization with 2-aminobenzoic acid. The analysis can be successfully performed within 15 min by using a CE instrument equipped with a He–Cd

LIF detector (Kamoda et al. 2006a, b). In Fig. 2, oligosaccharides derived from two serum samples are analyzed. The same set of oligosaccharides (**I, II, III, IV,** and **V**) is observed in both examples. However, relative abundances of oligosaccharides observed in myeloma IgG are obviously different from those in healthy subjects as indicated in the increase of the ratios of **IV** and **V**. These data may indicate the increase of the activity of galactosyl transferase.

Recently, we evaluated the performance of CE–LIF for the analysis of oligosaccharides derived from glycoprotein pharmaceuticals and demonstrated that excellent precision in migration times and peak areas of oligosaccharides (Kamoda et al. 2006a, b). CE provides detailed information on the oligosaccharide moieties in glycoprotein, and will contribute to the development of glycoprotein pharmaceuticals.

References

Kakehi K, Honda S (1996) Analysis of glycoproteins, glycopeptides and glycoprotein-derived oligosaccharides by high-performance capillary electrophoresis. J Chromatogr A 720(1–2):377–393

Kakehi K, Kinoshita M, Hayase S, Oda Y (1999) Capillary electrophoresis of N-acetylneuraminic acid polymers and hyaluronic acid: correlation between migration order reversal and biological functions. Anal Chem 71(8):1592–1596

Kakehi K, Kinoshita M, Nakano M (2002) Analysis of glycoproteins and the oligosaccharides thereof by high-performance capillary electrophoresis-significance in regulatory studies on biopharmaceutical products. Biomed Chromatogr 16(2):103–115

Kamoda S, Nakanishi Y, Kinoshita M, Ishikawa R, Kakehi K (2006a) Analysis of glycoprotein-derived oligosaccharides in glycoproteins detected on two-dimensional gel by capillary electrophoresis using on-line concentration method. J Chromatogr A 1106(1–2):67–74

Kamoda S, Ishikawa R, Kakehi K (2006b) Capillary electrophoresis with laser-induced fluorescence detection for detailed studies on N-linked oligosaccharide profile of therapeutic recombinant monoclonal antibodies. J Chromatogr A 1133(1–2):332–339

Section II
Sequence Analysis

Energy-Resolved Mass Spectrometry (ERMS) of Oligosaccharide

Ayako Kurimoto, Shusaku Daikoku, Osamu Kanie

Introduction

Mass spectrometry-based analytical method is extremely important in the elucidation of oligosaccharide structures because of its high sensitivity. Limited information regarding isomeric structure can be obtained by comparing certain fragments after CID experiments, and such comparisons can only be made when a set of isomers is available. Furthermore, quantitative analyses of MS/MS spectra have rarely been focused. In order to obtain reliable MS-based information, such analysis is inevitable. The energy-resolved mass spectrometry (ERMS) is thought to be promising in this regard.

The quadrupole ion-trap mass spectrometry (QIT-MS) provides low energy collision-induced dissociation process, has been used to resolve kinetic energy by obtaining a plot of the absolute abundance of precursor and product ions versus the amplitude of radio frequency (RF) applied on end cap electrodes. Recent advances in QIT-MS equipment allowed us to obtain CID spectra at MS^n. Thus, the ERMS can be performed at different stages of MS^n, which will be not only useful for the structural determination of a wide range of isomeric compounds but also important to obtain quantitative information regarding gas-phase reaction under CID conditions (Daikoku et al. 2007; Kurimoto et al. 2006).

Procedure

Instrumentation

Mass spectrometric experiments were performed on an Esquire 3000 plus (Bruker Daltonics GmbsH, Bremen, Germany), a quadrupole ion-trap mass spectrometer equipped with an ESI source (Nano-spray is preferable for the analysis of minute amounts of sample.). The samples were introduced into the ion source via infusion (flow rate: 120 μL/h). Only end cap RF amplitude was controlled during the CID experiment. He pressure: 4.86×10^{-6} mbar. CID time: 40 ms. Other parameters were as follows: dry temp: 250°C; Nebulizer gas (N_2): 10 psi, drying gas (N_2): 6.0 L/min, sample solutions: all samples were prepared in a mixed solution of MeOH/water (200 : 1) where concentrations varied in a range of pmol/mL. Smart frag: Off; scan range m/z 50~1,300; compound stability: 100%; ICC (ion charge control) target: 5000; max accumulation time: 200 ms. Other systems can be used as far as similar conditions were applied.

Mitsubishi Kagaku Institute of Life Sciences (MITILS), Minamiooya 11, Machida, Tokyo 194-8511, Japan
Phone: +81-42-724-6238, Fax: +81-42-724-6317
E-mail: kokanee@mitils.jp

Fig. 1 A typical ERMS for the pyridylaminated (PA-) maltohexasasac-
charide is shown

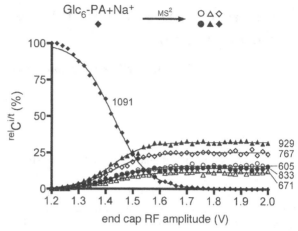

Data Collection

In our MS^n experiments, the supplementary dipolar RF voltage applied on end cap electrodes was raised by 0.02 V increments until the precursor ion could no longer be detected
(plateau at less than 0.9% of total ion current). Averages of n-4 spectra were used for
CID experiments ($n = 13{\sim}21$ where n equals number of scans); the first and the last two
datasets, which is associated with a transient period to steady state, in an RF amplitude
step were not used to avoid any inaccuracy. In the case that an intensity of a fragment
ion [FI] was greater than 5%, isotopic peaks with [FI + 1] and [FI + 2] were treated in
the calculations. For the isolation of a dater ion, $m/z \pm 2$ were isolated and subjected to
the CID experiments to include isotopes.

Data Handling

In order to obtain graphs of the energy-resolved mass spectra (Fig. 1), the following equa-
tions were used. When an ion A produces a series of product ions, $a, b, c \ldots, i, \ldots, n$,
the relative ion current for individual ions were defined by the equation,

$$^{rel}C^{i/t} = \frac{C^i}{C^A + \sum_a^n C^i} \times 100,$$ where $^{rel}C^{i/t}$ indicates ion current of a given ion among

observed ions in percentage, C^i is observed ion current in focus, C^A is ion current of a
precursor ion, and $^{rel}C^i$ is ion current of an ion in focus.

References

Daikoku S, Ako T, Kurimoto A, Kanie O (2007) Anomeric information obtained from a series of syn-
    thetic trisaccharides using energy-resolved mass spectra. J Mass Spectrom 42:714–723
Kurimoto A, Daikoku S, Mutsuga S, Kanie O (2006) Analysis of energy-resolved mass spectra at MS^n
    in a pursuit to characterize structural isomers of oligosaccharides. Anal Chem 78:3461–3466

LC/MS of N-Linked Oligosaccharides

Nana Kawasaki, Satsuki Itoh, Teruhide Yamaguchi

Introduction

Most glycoproteins are attached to some of diverse oligosaccharides differing in their monosaccharide compositions, sequences, branches, linkages, and modifications. Liquid chromatography/mass spectrometry (LC/MS) is recognized as a powerful tool for the separation and characterization of these oligosaccharides (Morelle et al. 2006; Zaia 2004). Here, we present the procedure for the LC/MS of N-linked oligosaccharides released from a glycoprotein.

Procedures

Sample Preparation

1. Dissolve a glycoprotein (50 µg) in 135 µL of 0.5 M Tris–HCl (pH 8.6) containing 8 M guanidine hydrochloride and 5 mM ethylenediamine tetraacetic acid.
2. Incubate the sample with 2-mercaptoethanol (1 µl) at 25°C for 2 h.
3. Add 2.8 mg of sodium monoiodoacetate and incubate the mixture at 25°C for 2 h in the dark.
4. Remove the excess reagent by gel filtration with Sephadex G-25 (e.g., PD-10 column, GE Healthcare), and lyophilize the eluant.
5. Dissolve the oligosaccharides in 50 µl of 0.1 M phosphate buffer (pH 7.2) and incubate the sample with N-Glycosidase F (PNGase F, 2 units) at 37°C for 16 h.
6. Add cold ethanol (125 µl) to precipitate the deglycosylated protein and maintain it at 4°C for 1 h.
7. Dry the supernatant after centrifugation at 15,000 rpm for 15 min.
8. Dissolve the oligosaccharides in water (100 µl), and add 0.5 M $NaBH_4$ (100 µl) to the solution.
9. Incubate the mixture at 25°C for 2 h.
10. Terminate the reaction with diluted acetic acid and, extract the borohydrate-reduced oligosaccharides with a solid-phase tip (e.g., a carbon tip, Supelco, Bellefonte, PA, USA).
11. Dry the eluant and dissolve the borohydrate-reduced oligosaccharides in water (10–50 µl).

Division of Biological Chemistry and Biologicals, National Institute of Health Sciences, 1-18-1 Kamiyoga, Setagaya-ku, Tokyo 158-8501, Japan
Phone: +81-3-3700-9074, Fax: +81-3-3700-9084
E-mail: nana@nihs.go.jp

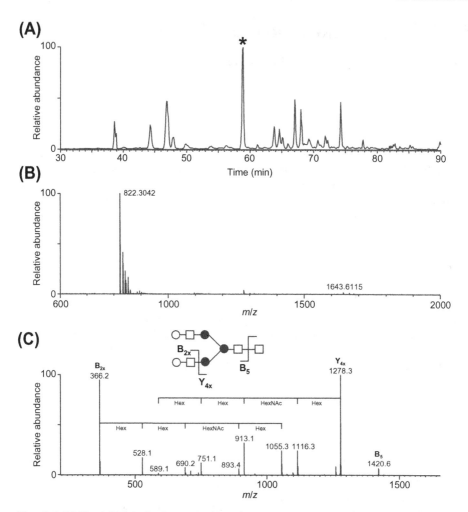

Fig. 1 LC/MS of *N*-linked oligosaccharides. (**A**) TIC obtained by LC/MS of oligosaccharides. (**B**) Typical mass spectrum acquired at 58.8 min (peak *). (**C**) Product ion spectrum acquired from *m/z* 822.3042 as a precursor. Sample: FSH. LC: column, 0.075 × 150 mm (Thermo Fisher Scientific, Waltham, MA, USA); flow rate, 200 nl/min. MS: LTQ-FT (Thermo Fisher Scientific)

LC/MS

1. Equilibrate the graphitized carbon column (0.075–0.2 mm i.d. × 150 mm) by running 95% of solvent A (5 mM ammonium acetate, 2% acetonitrile, pH 9.6) through it at an appropriate flow rate (0.2–200 μl/min) for 20–30 min.

2. Attach the nanospray tip to the outlet of the column, and set it on the x–y–z translational stage.

3. Set up the mass spectrometer (triple-stage quadrupole, ion trap, and quadrupole time of flight type analyzers are recommended) and check the stable spray with an elution buffer.

4. Inject samples (1–2 μl) onto the column and simultaneously start both the chromatography gradient and mass spectrometer data collection. The column is typically

eluted by a linear gradient from 5 to 50% of solvent B (5 mM ammonium acetate, 80% acetonitrile, pH 9.6) in 110 min. A single MS scan (m/z 600–2,000) followed by tandem mass spectrometry (MS/MS) should be acquired in a data-dependent manner.

Comment

It is possible to use in-gel carboxymethylation and in-gel PNGase F digestion for the release of oligosaccharides from a gel-separated glycoprotein (Itoh et al. 2006). As a typical example, we show the total ion chromatogram (TIC), mass spectrum and product ion spectrum of *N*-linked oligosaccharides released from recombinant human follicle stimulating hormone (FSH) separated by SDS-PAGE (Fig. 1).

References

Itoh S, Kawasaki N, Hashii N, Harazono A, Matsuishi Y, Hayakawa T, Kawanishi T (2006) *N*-linked oligosaccharide analysis of rat brain Thy-1 by liquid chromatography with graphitized carbon column/ion trap-Fourier transform ion cyclotron resonance mass spectrometry in positive and negative ion modes. J Chromatogr A 1103:296–306

Morelle W, Canis K, Chirat F, Faid V, Michalski JC (2006) The use of mass spectrometry for the proteomic analysis of glycosylation. Proteomics 6:3993–4015

Zaia J (2004) Mass spectrometry of oligosaccharides. Mass Spectrom Rev 23:161–227

Convenient Structural Characterization of Intact Glycosphingolipids by MALDI-TOF Mass Spectrometry with Increased Laser Power and Cooling Gas Flow

Minoru Suzuki

Introduction

Matrix-assisted laser desorption/ionization quadrupole ion trap time-of-flight mass spectrometry (MALDI-QIT-TOF MS) was applied to the structural characterization of intact glycosphingolipids. Lithium adduct ions of glycosphingolipids were analyzed using MALDI-QIT-TOF MS (Suzuki et al. 2005) under strong conditions of increased laser power and cooling gas flow. Consequently, the fragmentation patterns of MS and MS^2 for the glycosphingolipids under the strong conditions were comparable to those of MS^2 and MS^3 obtained under standard conditions, respectively (Suzuki et al. 2006).

Results

Because lithium adduct ions are more informative than sodium adduct ions for the structural characterization of ceramides, lithium adduct ions were generated by the addition of LiCl to the intact glycosphingolipid solutions before analysis. In the MALDI-TOF MS spectrum of Forssman glycolipid obtained under the standard conditions, the lithiated molecular ion was detected at m/z 1,548, and no other fragment ions were observed. In the MS^2 spectrum of Forssman glycolipid with selection of the ion at m/z 1,548, fragment ions due to the elimination of GalNAc (m/z 1,345), GalNAc-GalNAc (m/z 1,142), GalNAc-GalNAc-Gal (m/z 980), GalNAc-GalNAc-Gal-Gal (m/z 818), and GalNAc-GalNAc-Gal-Gal-Glc (m/z 656). The [GalNAc-GalNAc-Gal + Li] (m/z 575) and GalNAc-GalNAc + Li] (m/z 413) ions were also detected (Fig. 1a). In the MS^3 spectrum selecting the LacCer ion (m/z 980), the most abundant ion detected by MS^2 was performed for the structural characterization of the ceramide moiety. In the MS^3 spectrum, the ions resulting from the elimination of Gal (m/z 818) and Gal-Glc (m/z 656) were detected, as well as the d18-sphingenine at 289 (Fig. 1b). MALDI-TOF MS spectrum and MS^2 spectrum of Forssman glycolipid were obtained with increased laser power and cooling gas flow; as a result, the fragmentation patterns of MS and MS^2 for Forssman glycolipid under the strong conditions were comparable to those of MS^2 and MS^3 obtained under standard conditions, respectively (Fig. 2). Thus, MALDI-TOF MS with the strong conditions is a convenient method for the intact glycosphingolipid analysis.

Sphingolipid Expression Laboratory, RIKEN Frontier Research System, 2-1 Hirosawa, Wako, Saitama 351-1098, Japan
Phone: +81-48-467-4404, Fax: +81-48-462-4615
E-mail: m-suzuki@riken.jp

Fig. 1 MALDI-QIT-TOF MS² and MS³ spectra of Forssman glycolipid obtained under the standard conditions

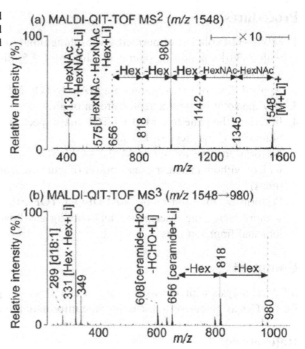

Fig. 2 MALDI-QIT-TOF MS and MS² spectra of Forssman glycolipid obtained under the strong conditions

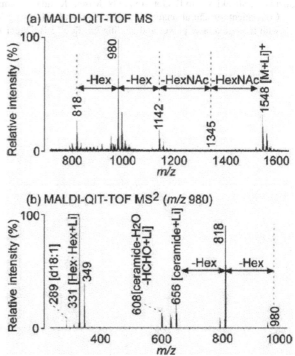

Procedures

1. Mix 2–3 μl solution of purified glycosphingolipid in chloroform-methanol (2 : 1, v/v) with 2–3 μl solution of saturated solution of 2,5-dihydroxy benzoic acid in 40 mM aqueous LiCl used as the matrix.
2. Apply 1–2 μl of the solution on a MALDI-TOF MS plate.
3. Dry the solution with a gentle stream of air.
4. Introduce the plate to a MALDI-TOF mass spectrometer (Axima MALDI-QIT-TOF, Shimadzu Corp., Kyoto, Japan).
5. Perform positive ion mode MALDI-QIT-TOF MS with a 337 nm pulsed N2 laser, with or without increasing laser power (about twice) and cooling gas flow (about 1.5–2 times).
6. Perform positive ion mode MALDI-QIT-TOF MS/MS and/or MS/MS/MS with or without increasing laser power and cooling gas flow, selecting Li adduct molecular ions and fragment ions.

Comment

MS/MS analysis with increased laser power and cooling gas, selecting an ion responsible for LacCer as precursor ion, can provide information on ceramide structure.

References

Suzuki Y, Suzuki M, Ito E, Ishii H, Miseki K, Suzuki A (2005) Convenient and rapid analysis of linkage isomers of fucose-containing oligosaccharides by matrix-assisted laser desorption/ionization quadrupole ion trap time-of-flight mass spectrometry. Glycoconjugate J 22:427–431

Suzuki Y, Suzuki M, Ito E, Goto-Inoue N, Miseki K, Iida J, Yamazaki Y, Yamada M, Suzuki A (2006) Convenient structural analysis of glycosphingolipids using MALDI-QIT-TOF mass spectrometry with increased laser power and cooling gas flow. J Boichem 139:771–777

Structural Analyses of Glycoconjugates by NMR

Koichi Kato, Yoshiki Yamaguchi

Introduction

NMR spectroscopy has been one of the most powerful methods to obtain structural information of oligosaccharides. Vliegenthart and co-workers have greatly achieved systematic studies on ^{1}H chemical shifts of oligosaccharides and thereby described numerous structures by using a "structural reporter group" (Vliegenthart 1980). Recent advances in NMR and stable isotope labeling techniques have made possible not only determination of covalent structure of oligosaccharides but also analyses of conformations and dynamics of glycoconjugates, which are usually inaccessible by X-ray crystallography.

Concept

NMR analyses of glycans have been hampered by spectral overlap and low sensitivity. Recently, NMR sensitivity has been greatly improved by the emergence of higher magnetic fields and cryogenic probes. Higher magnetic fields also offer higher resolution spectra containing sufficient information for detailed structural analyses of complicated oligosaccharides (Fig. 1).

In NMR spectra of oligosaccharides, chemical shifts and spin–spin coupling constants provide information on the positions and configurations of glycosidic linkage. ^{1}H and ^{13}C NMR chemical shift database is available at the websites, GLYCOSCIENCES.de (http://www.glycosciences.de/) and SUGABASE (http://www.boc.chem.uu.nl/sugabase/sugabase.html). For determination of the glycosidic conformation, nuclear Overhauser effects (NOEs) are generally used, which are related to the inverse six power of the ^{1}H–^{1}H distances. However, collection of enough number of conformational restraints is so difficult that glycosidic torsion angles have often been described based on NMR data combined with molecular dynamics simulations (Woods et al. 1998). At higher magnetic fields, it becomes easier to make oligosaccharide molecules undergo anisotropic rotational Brownian motion, which enables us to observe residual dipolar couplings (RDCs). RDC can be a useful NMR parameter for determining the conformation of oligosaccharides, because its magnitude depends on the angle between the principal axis of the alignment tensor and the bond vector connecting the two nuclei such as pair of ^{13}C and ^{1}H at the CH group (Prestegard et al. 2004).

^{13}C-labeling of oligosaccharides provides an avenue for conformational analysis using $^3J_{CH}$ and $^3J_{CC}$ coupling constants, which define the torsion angles more unambiguously (Duker and Serianni 1993). We have been developing a systematic strategy for stable

Graduate School of Pharmaceutical Sciences, Nagoya City University, Tanabe-dori 3-1,
Mizuho-ku, Nagoya 467-8603, Japan
Phone: +81-52-836-3447, Fax: +81-52-836-3447
E-mail: kkato@phar.nagoya-cu.ac.jp

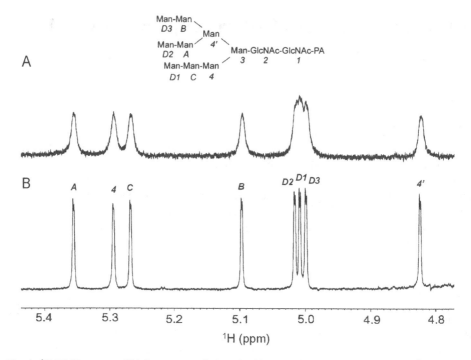

Fig. 1 ^{1}H-NMR spectra of high-mannose oligosaccharide (Man$_9$GlcNAc$_2$) recorded at the ^{1}H observation frequencies of **A** 500 MHz and **B** 920 MHz

isotope labeling for detailed NMR studies of glycoproteins in solution (Yamaguchi and Kato 2006). In this methodology, the glycans and/or polypeptides of glycoproteins are uniformly or selectively labeled with stable isotopes (^{13}C, ^{15}N, and ^{2}H) by metabolic or enzymatic manners (Fig. 2). NMR spectral data obtained by use of the isotopically labeled glycoprotein provide us with information of structure and dynamics of the carbohydrate moieties attached in situ to proteins at atomic resolution.

Procedures of ^{13}C and ^{15}N Metabolic Labeling of Glycoproteins for NMR Study

We herein present a protocol for ^{13}C and ^{15}N labeling experiment of the GlcNAc residues of carbohydrate chains expressed by IgG-Fc. The procedure subsequently described has been established for mouse hybridoma cells producing monoclonal IgG. Some modifications might be necessary for other types of cells.

1. Prepare a modified Nissui NYSF 404 medium containing 0.2 g/L of [^{15}N]GlcN·HCl and 1 g/L of [^{13}C$_6$]glucose (See Table 1).
2. Adapt the hybridoma cells to the serum-free medium and cultivate the cells in 4 L of the medium for 2 weeks.
3. After cell growth, concentrate the culture supernatant with a Pellicon ultrafiltration system (Millipore) and then apply to an Affi-Gel protein A column (Bio Rad).
4. Digest the purified IgG with papain according to the literature (Yamaguchi et al. 1995).

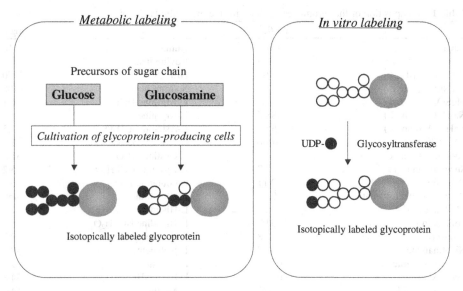

Fig. 2 A scheme of stable isotope labeling of glycoproteins

5. Purify the IgG-Fc with an Affi-Gel protein A column.
6. Concentrate the purified IgG-Fc solutions with Amicon Ultra (Millipore) to a final volume of 0.5 ml at a final glycoprotein concentration of 0.2 mM, in 5 mM sodium phosphate buffer, pH 6.0 containing 200 mM NaCl and 3 mM NaN$_3$ in 90% H$_2$O/10% D$_2$O.
7. Perform the heteronuclear NMR measurements of IgG-Fc (Fig. 3).

Under this culture condition, a majority of the IgG-Fc glycan is produced as agalactosylated forms due to the presence of GlcN in the medium.

Procedures of Introduction of ^{13}C-Galactose onto Glycoprotein for NMR Study

In vitro incorporation of isotopically labeled sugars into carbohydrate chains is another useful method to achieve stable isotope labeling of glycoproteins for NMR analyses (Yamaguchi and Kato 2006). A procedure for the enzymatic attachment of ^{13}C-galactose onto IgG-Fc glycoprotein as an example is presented.

Removal of Galactose Residues from IgG-Fc Glycoprotein

Fc fragment is cleaved from IgG by papain digestion (Yamaguchi et al. 1995). The galactose (Gal) residues of IgG-Fc glycoprotein should be removed prior to galactosylation by β-galactosidase treatment.

1. Dissolve the IgG-Fc (2 mg/mL) in 50 mM sodium acetate buffer, pH 5.5.
2. Add 0.02 units/mL of β-galactosidase (from *Streptococcus* 6646K, Seikagaku Corp.) with 10 mM MnCl$_2$.
3. Incubate at 37°C for 72 h.
4. Neutralize the reaction mixture with 1.5 M Tris–HCl, pH 8.5.

Table 1 Composition of the serum-free medium for the metabolic labeling[a]

	mg/L		mg/L
NaCl	6208	Human transferrin	10
KCl	388	Bovine insulin	10
$CaCl_2$ (unhyd.)	97	HEPES	3570
$Ca(NO_3)_2$ (unhyd.)	33.7	Sodium bicarbonate	1400
$MgSO_4$	69	L-Arginine·HCl	76.1
Na_2HPO_4 (unhyd.)	388.5	L-Arginine	97
NaH_2PO_4 (unhyd.)	55.8	L-Asparagine·H_2O	42.5
$[^{13}C_6]$Glucose	1000	L-Aspartic acid	9.7
Sodium succinate	97	L-Cystine·$2H_2O$	31.5
Succinic acid	72.8	L-Cysteine·HCl·H_2O	15.2
D-Biotin	0.11	L-Glutamic acid	9.7
D-Calcium pantothenate	0.61	L-Glutamine	450
Choline chloride	26.5	L-Histidine	7.3
Folic acid	0.97	L-Histidine·HCl·H_2O	20.4
i-Inositol	18	L-Hydroxyproline	9.7
Nicotinamide	0.97	L-Isoleucine	49.5
p-Aminobenzoic acid	0.485	L-Leucine	49.5
Pyridoxal·HCl	0.485	L-Lysine·HCl	54.8
Pyridoxine·HCl	0.485	L-Methionine	14.6
Riboflavin	0.146	L-Proline	14.7
Thiamin·HCl	0.97	L-Serine	29.6
Vitamin B_{12}	0.0037	L-Threonine	48
Glutathione (red.)	0.485	L-Valine	47
Choline bitartrate	0.873	L-Phenylalanine	22.8
Putrescine·$2H_2O$	0.0125	L-Tryptophan	7.3
Sodium pyruvate	220	L-Tyrosine	27.2
Thymidine	0.0125	Glycine	9.9
Hypoxanthine	0.025	$[^{15}N]$GlcN·HCl	200
Sodium selenite	0.0017		
Penicillin G	63.29		
Streptomycin	100		
Phenol red	5		

[a]The medium can be supplemented with 2 % (v/v) dialyzed fetal bovine serum.

5. Purify IgG-Fc from the reaction mixture using an Affi-gel protein A column (Bio-Rad Laboratories).
6. Check the degalactosylation by the HPLC mapping method.

Attachment of ^{13}C-Galactose onto IgG-Fc Glycoprotein

While commercially available UDP-Gal could be used for galactosylation with unlabeled D-Gal, it is necessary to generate UDP-$[^{13}C_6]$Gal in situ in the reaction mixture for galactosylation with D-$[^{13}C_6]$Gal.

1. Dissolve the degalactosylated Fc at a concentration of 2 mg/ml in 100 mM HEPES buffer, pH 7.4, containing 20 mM KCl, 5 mM $MgCl_2$, 10 mM ATP (Sigma), 4 units/ml of galactokinase (Sigma), and 10 mM D-$[^{13}C_6]$Gal.
2. Incubate at 37°C for 30 min.

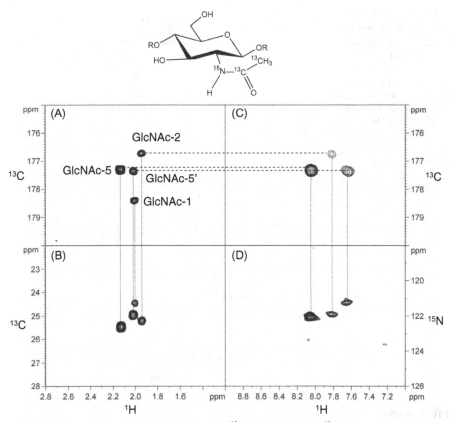

Fig. 3 2D NMR spectra of IgG-Fc labeled with [^{15}N]GlcN·HCl and [^{13}C$_6$]glucose. **A** 2D H(C)CO, **B** 2D HC(CO), **C** 2D H(N)CO and **D** ^{1}H-^{15}N HSQC spectra of IgG-Fc, which exhibit peaks derived from the acetamide groups of the GlcNAc residues. The NH peak from the innermost GlcNAc residue is not observed due to line broadening

3. Add 0.4 units/ml of Gal-1-phosphate uridyltransferase (Sigma) and 10 mM UDP-Glc (Sigma).
4. Incubate at 37°C for 4 h.
5. Add 5 mM MnCl$_2$, 2 units/ml of galactosyltransferase (Sigma), 20 mM potassium phosphoenolpyruvate (Sigma), 40 units/ml of pyruvate kinase (Sigma), 2 units/ml of inorganic pyrophosphatase (Sigma), and 2 units/ml of UDP-Glc pyrophosphorylase (Sigma).
6. Incubate at 37°C for 72 h.
7. Neutralize the reaction mixture with 1.5 M Tris–HCl, pH 8.5.
8. Purify IgG-Fc using an Affi-gel protein A column (Bio-Rad Laboratories).
9. Check the galactosylation by the HPLC mapping method.

An almost complete (>95%) galactosylation of IgG-Fc glycoprotein could be attained by this procedure. NMR spectrum of [^{13}C$_6$]Gal-labeled Fc is shown in Fig. 4.

Fig. 4 2D ^{1}H-^{13}C HSQC spectrum of Fc labeled with [^{13}C$_6$]Gal. The assignments were performed by 2D HCCH-COSY experiment combined with the branch-selective ^{13}C labeling method (Yamaguchi and Kato 2006). It is of note that Gal-6′ showed broader signals than Gal-6, indicating that the mobility of Gal-6′ is restricted in the Fc glycoprotein

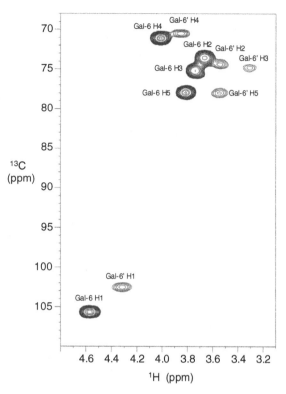

References

Duker JM, Serianni AS (1993) (^{13}C)-substituted sucrose: ^{13}C-^{1}H and ^{13}C-^{13}C spin coupling constants to assess furanose ring and glycosidic bond conformations in aqueous solution. Carbohydr Res 249:281–303

Prestegard JH, Bougault CM, Kishore AI (2004) Residual dipolar couplings in structure determination of biomolecules. Chem Rev 104:3519–3540

Vliegenthart JF (1980) High resolution ^{1}H-NMR spectroscopy of carbohydrate structures. Adv Exp Med Biol 125:77–91

Woods RJ, Pathiaseril A, Wormald MR, Edge CJ, Dwek RA (1998) The high degree of internal flexibility observed for an oligomannose oligosaccharide does not alter the overall topology of the molecule. Eur J Biochem 258:372–386

Yamaguchi Y, Kato K (2006) Structural glycobiology by stable-isotope-assisted NMR spectroscopy. In: Webb GM (ed) Modern magnetic resonance, vol 1. Springer, The Netherland, pp 219–225

Yamaguchi Y, Kim H, Kato K, Masuda K, Shimada I, Arata Y (1995) Proteolytic fragmentation with high specificity of mouse immunoglobulin G. Mapping of proteolytic cleavage sites in the hinge region. J Immunol Methods 181:259–267

Sugar Chain Analysis by Enzymatic Digestion and 2D Mapping by HPLC

Megumi Maeda[1], Yoshinobu Kimura[2]

Introduction

Structural analysis is crucial to elucidating the physiological functions of oligosaccharides. The rapid development of both hardware and software for NMR spectrometry and mass spectrometry (ESI-MS and MALDI-MS) has enabled significantly faster and easier structural analysis. When analyzing a small amount of oligosaccharide, the combined use of endo- and exo-glycosidases having clear-cut substrate specificities provides large and reliable advantages for structural analysis. For structural analysis using glycosidases, which are used for high-sensitivity detection, oligosaccharides are usually labeled with tritium or fluorescence reagents: 2-aminopyridine (Hase et al. 1978, 1987) or 2-amino-benzamide (Rudd et al. 1997) in advance. When fluorescent reagents are used for labeling oligosaccharides, reverse-phase HPLC can be used for structural analysis because of the increase in the hydrophobicity of sugar chains. Therefore, two-dimensional (2D)-HPLC using both reverse-phase and normal-phase columns can be used for identifying oligosaccharide structures. 2D-HPLC can distinguish the structural difference (anomeric or branching structures) between isomeric oligosaccharides of the same molecular size or sugar composition using a small amount of sample. Here, as a typical example, we describe the structural analysis of a novel N-glycan obtained from royal jelly glycoproteins using fluorescence labeling and 2D-HPLC developed by Hase et al. (1978, 1987).

Structural Analysis Using 2D-HPLC and Glycosidase Digestion

Royal jelly is a famous health food containing many bioactive glycoproteins that can stimulate the proliferation of human monocytic cell lines. Recently, we have identified novel N-glycans harboring a tumor-related antigen (T antigen) on royal jelly glycoproteins (Kimura et al. 2006). The N-glycans were liberated from glycopeptides by hydrazinolysis, and the resulting free oligosaccharides were labeled with 2-aminopyridine. The resulting pyridylaminated (PA) sugar chains (more than 20) were purified by a combination of reverse-phase (RP)-HPLC and size-fractionation (SF)-HPLC, and the structure of each sugar chain was analyzed preliminarily by 2D-HPLC and ESI-MS. The structures of the major N-glycans (Man9~4GlcNAc2, GlcNAc2Man3GlcNAc2) were identified by

[1]Department of Hygiene, Kawasaki Medical School, 577 Matsushima, Kurashiki 701-0192, Japan

[2]Department of Biofunctional Chemistry, Graduate School of Natural Science and Technology, Okayama University, 1-1 Tsushima-Naka, Okayama 700-8530, Japan
Phone: +81-86-251-8296, Fax: +81-86-251-8388
E-mail: yosh8mar@cc.okayama-u.ac.jp

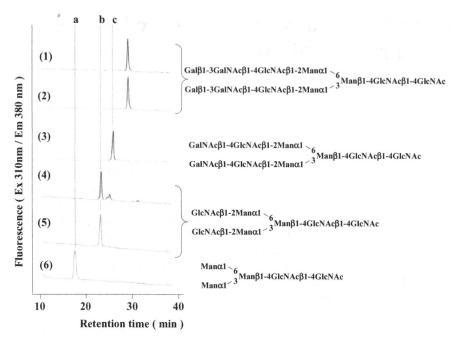

Fig. 1 SF-HPLC of sequential glycosidase digests of PA-sugar chain. *1* Royal jelly *N*-glycan (HexNAc5Hex5GlcNAc-PA); *2* jack bean β-*N*-acetylhexosaminidase digest of 1; *3* β1–3 galactosidase digest of 1; *4* β-*N*-acetylgalactosaminidase digest of 3; *5 Streptomyces* lacto-*N*-biosidase digest of 1; *6* diplococcal β-*N*-acetylhexosaminidase digest of 5. *a*, *b*, and *c* indicate the elution positions of the authentic PA-sugar chains, Man3GlcNAc2-PA, GlcNAc2Man3GlcNAc2-PA, and GalNAc2GlcNAc2Man3GlcNAc2-PA, respectively

a combination of 2D PA-sugar chain mapping and ^{1}H-NMR. However, the elution positions of some complex-type *N*-glycans did not correspond with those of any available authentic PA-sugar chains on 2D-HPLC. ESI-MS revealed that one of the complex-type *N*-glycans consisted of *N*-acetylhexosamine (5 residues), hexose (5 residues), and PA-GlcNAc (HexNAc5Hex5GlcNAc-PA). Jack bean β-*N*-acetylhexosaminidase released no HexNAc residues, suggesting that the nonreducing end sugar residue was not *N*-acetylhexosamine (Fig. 1(2)). On the other hand, recombinant β1-3 galactosidase released galactose residues from HexNAc5Hex5GlcNAc-PA, suggesting the presence of β1-3 galactosyl residues at the nonreducing end (Fig. 1(3)). Then, the product was converted to GlcNAc2Man3GlcNAc2-PA by β-*N*-acetylgalactosaminidase digestion, and the elution position on 2D-HPLC corresponded with that of GlcNAcβ1-2Manα1-6(GlcNAcβ1-2Manα1-3)Manβ1-4GlcNAcβ1-4GlcNAc2-PA, suggesting that the two GalNAc residues are linked to GlcNAc by β-linkage (Fig. 1(4)). *Streptomyces* lacto-*N*-biosidase, which can release the Galβ1-3HexNAc unit but not the Galβ1-4HexNAc unit from complex-type *N*-glycans, converted the starting *N*-glycan (HexNAc5Hex5GlcNAc-PA) to GlcNAc2Man3GlcNAc2-PA, indicating the presence of two Galβ1-3GalNAc units (Fig. 1(5)). The lacto-*N*-biosidase digest was further converted to the trimannosyl core structure, Man3GlcNAc2-PA, by diplococcal β1-2-specific β-*N*-acetylglucosaminidase digestion (Fig. 1(6)). From these results of sequential glycosidase digestion, a unique *N*-glycan structure with two T antigen (Galβ1-3GalNAc) units was proposed as shown in Fig. 1 (uppermost structure). Furthermore, the novel structure deduced by sequential glycosidase digestions was further confirmed by methylation and ^{1}H-NMR analyses (Kimura et al. 2006).

Thus, 2D-HPLC using fluorescence-labeled *N*-glycans enables the determination or deduction of *N*-glycan structures using a small amount of the sample. However, note that the combined analysis using well-characterized glycosidases seems to be essential for the precise identification of their structures.

Protocol for Glycosidase Digestion of PA-sugar Chain

Sequential glycosidase digests were analyzed by HPLC using a Shodex Asahipak NH2P-50 column. Man3GlcNAc2-PA, GlcNAc2Man3GlcNAc2-PA, and GalNAc2GlcNAc2-Man3GlcNAc2-PA were used as the authentic PA-sugar chains.

1. Incubate 200 pmol PA-sugar chain (HexNAc5Hex5GlcNAc-PA) with 600 mU Jack bean β-*N*-acetylhexosaminidase in 20 μL of 0.1 M sodium acetate buffer, pH 4.0, for 2 days at 37°C.
2. Incubate 200 pmol PA-sugar chain (HexNAc5Hex5GlcNAc-PA) with 190 mU β1-3 specific galactosidase in 20 μL of 0.1 M sodium acetate buffer, pH 5.5, at 37°C for 16 h.
3. Incubate the resulting product of step 2 with 220 mU *Bacillus* β-*N*-acetylgalactosaminidase in 20 μL of 0.1 M sodium acetate buffer, pH 6.5, at 37°C for 16 h.
4. Incubate 200 pmol PA-sugar chain (HexNAc5Hex5GlcNAc-PA) with 1 μU *Streptomyces* lacto-*N*-biosidase in 20 μL of 0.1 M sodium acetate buffer, pH 5.5, for 2 days at 37°C.
5. Incubate the resulting product of step 4 with 10 mU diplococcal β-*N*-acetylhexosaminidase in 20 μL of 0.1 M sodium acetate buffer, pH 4.8, for 2 days at 37°C.
6. Stop these reactions by boiling at 100°C for 3 min.
7. Centrifuge at 15,000 rpm for 10 min.
8. Inject 20 μl out of the mixture of 70% acetonitrile (65 μL) and 5 μl of the resulting supernatant of the reaction mixture into size-fractionation (SF) HPLC.
9. Elute these PA-sugar chains by increasing the water content in the water–acetonitrile mixture from 26 to 50% linearly for 40 min at a flow rate of 0.7 ml/min.
10. For detection of PA-sugar chains, an excitation wavelength of 310 nm and an emission wavelength of 380 nm were used.

References

Hase S, Ikenaka T, Matsushima Y (1978) Structure analyses of oligosaccharides by tagging of the reducing end sugars with a fluorescent compound. Biochem Biophys Res Commun 85:257–263

Hase S, Natsuka S, Oku H, Ikenaka T (1987) Identification method for twelve oligomannose-type sugar chains thought to be processing intermediates of glycoproteins. Anal Biochem 167:321–326

Kimura Y, Ushijima T, Maeda M, Hama Y, Kimura M, Okihara K, Sugimoto H, Yamada H (2006) Tumor antigen occurs in *N*-glycan of royal jellyglycoproteins: Honeybee cells synthesize T-antigen unit in *N*-glycan moiety. Biosci Biotechnol Biochem 70:2583–2587

Rudd PM, Guile GR, Küster B, Harvey DJ, Opdenakker G, Dwek A (1997) Oligosaccharide sequencing technology. Nature 388:205–207

Analysis of Binding Sites of Sugar Chains by Methylation Analysis

Sumihiro Hase

Introduction

The linkage structures between the monomer units in most proteins and nucleic acids are one of a kind; however, for carbohydrates the linkage structures between the monosaccharide units, acetal or ketal linkage (glycosidic linkage), are more than one of a kind, including the anomeric configuration and the linkage positions of available OH groups. Therefore, determination of the linkage positions of the monosaccharide residues is inevitable for determination of the chemical structures of sugar chains. This analysis may be partly done by Smith degradation, mass spectrometry, hydrolysis with specific enzymes, and NMR; however, methylation analysis is the most reliable method in this field especially for determination of novel sugar structures.

Results

The principle of the method includes complete methylation of free OH groups under the alkaline reaction conditions, and then the products are hydrolyzed to partially methylated monosaccharides, which are reduced to sugar alcohols followed by peracetylation. Partially methylated alditol acetates thus obtained are separated and analyzed by gas-liquid chromatography-mass spectrometry (GC-MASS) using electron impact ionization or chemical ionization. Based on the fragment patterns and the retention times, the positions of non-methylated hydroxy groups are assigned. Identification by HPLC of partially methylated alditol acetates is also done as pyridylamino derivatives followed by HPLC for micro-quantities of a sample (Wakabayashi et al. 1999). Methylation analysis is the standard method to determine the linkage positions; however, this largely depends on the degree of the methylation reaction. The procedure for complete methylation was largely improved when soluble sodium methylsulfinyl carbanion was introduced as a strong base instead of solid bases (Hakomori 1964). This method has been improved and became a standard method of complete methylation. Later lithium methylsulfinyl carbanion (Parente et al. 1985; Blakeney and Stone 1985) and sodium hydroxide powder (Ciucanu and Kerek 1984) were introduced as bases for micro-quantities of glycoconjugates due to a lesser background. However, methylation analysis recently tends to be avoided, as informations of the linkage positions are partly done by NMR, mass spectrometry, and digestion with specific enzymes. However, methylation analysis is still needed for determining unknown sugar structures. Therefore, highly sensitive and automated procedure is awaited. Readers

Department of Chemistry, Graduate School of Science, Osaka University, 1-1
Machikaneyama-cho, Toyonaka, Osaka 560-0043, Japan
Phone: +81-6-6850-5380, Fax: +81-6-6850-5382
E-mail: suhase@chem.sci.osaka-u.ac.jp

are referred to the excellent published reviews (Geyer et al. 1983; Geyer and Geyer 1994).

References

Blakeney AB, Stone BA (1985) Methylation of carbohydrates with lithium methylsulphinyl carbanion. Carbohydr Res 140:319–324

Ciucanu I, Kerek F (1984) A simple and rapid method for the permethylation of carbohydrates. Carbohydr Res 131:209–217

Geyer R, Geyer H, Kühnhardt S, Mink W, Stirm S (1983) Methylation analysis of complex carbohydrates in small amounts: capillary gas chromatography-mass fragmentography of methylalditol acetates obtained from N-glycosidically linked glycoprotein oligosaccharides. Anal Biochem 133:197–207

Geyer R, Geyer H (1994) Saccharide linkage analysis using methylation and other techniques. In: Lennarz WJ, Hart GW (eds) Methods in enzymology, vol 230. Academic Press, San Diego, pp 86–108

Hakomori S (1964) A rapid permethylation of glycolipid, and polysaccharide catalyzed by methylsulfinyl carbanion in dimethyl sulfoxide. J Biochem 55:205–208

Parente JP, Cardon P, Leroy Y, Montreuil J, Fournet B (1985) A convenient method for methylation of glycoprotein glycans in small amounts by using lithium methylsulfinyl carbanion. Carbohydr Res 141:41–47

Wakabayashi H, Natsuka S, Mega T, Otsuki N, Isaji M, Naotsuka M, Koyama S, Kanamori T, Sakai K, Hase S (1999) Novel proteoglycan linkage tetrasaccharides of human urinary soluble thrombomodulin, SO_4-3GlcAβ1-3Galβ1-3($\pm$Siaα2-6)Galβ1-4Xyl. J Biol Chem 274:5436–5442

Glycan Profiling

Jun Hirabayashi

Introduction

Glycan profiling is a novel approach to structural analysis of glycans. Its emphasis is on grasping essential information of target glycans in a rapid, sensitive, and high-throughput manner rather than defining covalent structures in a rigorous and time-consuming manner. This information is particularly important for structural glycomics with the absence of a useful automated sequencer analogous to protein and DNA sequencers, which are based on Edman degradation chemistry and Sanger's dideoxy method, respectively. However, it is obviously difficult to apply this type of logic to glycans, because they have additional aspects increasing structural variety, which the linear molecules lack. They are represented by linkage isomers (i.e., α/β1-2, 1-3, 1-4, 1-6) and branching events. For this type of branching molecules, it seems that a "memory-matching" or "finger-printing" principle should work better. In this chapter, essence of practical approaches to glycan profiling is described.

Approaches to Glycan Profiling

To achieve glycan profiling, there are three types of separation principles based on "chemistry (LC, liquid chromatography)", "physics (MS, mass spectrometry)", and "biology (lectin affinity)". Evidently, all of these principles are independent from one another, and thus, form a principle "triad" (Fig. 1) to compensate the other's defects in a most efficient way (Hirabayashi 2004). From a practical viewpoint, glycan profiling is carried out in two different ways: one is to characterize glycans liberated from glycoproteins, cells and tissues (designated here "liberated glycomics"), and the other is to directly target glycoproteins, or those extracted from cells and tissues ("intact glycomics").

Liberated glycomics is a general method for glycan profiling. In this method, glycans are liberated from the core protein prior to analysis by either chemical (e.g., hydrazinolysis) or enzymatic (e.g., peptide-*N*-glycanase digestion) procedures, and the liberated glycans are, in general, subjected to fluorescence labeling for detection. In this respect, a special equipment has been developed for release and catch of glycans from glycoproteins (Shimaoka et al. 2007). In liberated glycomics, however, any information of core proteins, to which glycans originally attached, as well as that of attachment sites, is lost by the liberation procedure. On the other hand, intact glycomics is simple and straightforward. Probably, the latter approach is more suitable for rapid analysis of various

Lectin Application and Analysis Team, Research Center for Medical Glycoscience, National Institute of Advanced Industrial Science and Technology, AIST, Tsukuba Central 2, 1-1-1, Umezono, Tsukuba, Ibaraki 305-8568, Japan
Phone: +81-29-861-3124, Fax: +81-29-861-3125
E-mail: jun-hirabayashi@aist.go.jp

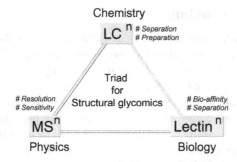

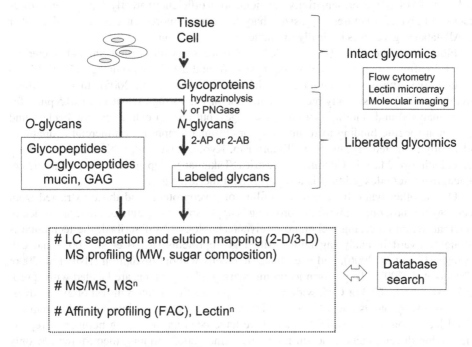

Fig. 1 A principle "triad" contributing to practical approaches to glycan profiling. Three strategies based on different separation principles, i.e., chemistry, physics, and biology, are compensatory to one another, and thus their combinations work efficiently in various fields of glycomics

Fig. 2 A general scheme for glycan profiling

glycoprotein drugs (i.e., quality control) including antibody, whereas at the moment more precise (rigorous) information of glycan structures is obtained only by liberated glycomic procedures. In order to clarify biological roles of glycans, e.g., in glycan-primed bio-signaling events, glycan profiling should be made in an intact cell/tissue state without destruction of cell surface structures. However, such approaches, if possible, are extremely limited and laborious. These may include flow-cytometry analysis using anti-glycan antibodies and lectins. More recently, intact cell/tissue profiling methods have been developed as advanced techniques of MS and lectin microarray. In future, glycan-targeted molecular imaging might be realized to analyze glycan localization, assembled features as well as their covalent structures. A general scheme of glycan profiling is summarized in Fig. 2.

Practice of Glycan Profiling

As described, prior fluorescence labeling is essential to perform liberated glycomics not only to increase sensitivity in separation but also to enhance separation resolution in HPLC. That is why relatively small (MW, ca. 100) fluorescence reagents, such as 2-aminopyridine (2-PA) and 2-aminobenzamide (2-AB) have been widely used for multi-dimensional LC mapping (i.e., chemical mapping), because their relatively weak hydrophobicity assures rapid dissociation from the chromatography resin, and thus, provide sharp elution peaks. Fluorescence labeling is also effective in structural analysis by mass spectroscopy, because it substantially increases ionization efficiency. Among MS technologies, tandem-mass spectrometry analysis called MS/MS or MS^n is a powerful method to differentiate resembling glycan structures (e.g., isomers) in a spectral-matching manner (i.e., physical mapping). In both cases, required amounts of glycans are in the order of pmol or less. As a defect of MS, however, sensitivity (i.e., ionization efficiency) greatly depends on various factors of glycan structures (e.g., size, charge), whereas fluorescence intensity of 2-AP or 2-AB-labeled glycans is basically the same in LC separation.

Fluorescence labeling (2-AP or 2-AB) is also necessary for lectin-affinity characterization (bio-affinity mapping) and works best in frontal affinity chromatography (FAC), a quantitative affinity chromatography (Nakamura-Tsuruta et al. 2005). In this method, pyridylamination is strongly recommended not only in its high sensitivity (<0.5 pmol for each analysis) and stability (long-term storage), but also in extremely low background (i.e., non-specific binding to resins used for chromatography). Commercial availability of pyridylaminated glycans (e.g., Takara Bio, Kyoto, Japan, http://bio.takara.co.jp/bio_en/default.asp; Masuda Chemical Industries, Takamatsu, Japan, http://www.mc-ind.co.jp/english_site/index_e.html) is also a practical merit for users.

On the other hand, for a direct profiling of glycoproteins and those extracted from cells and tissues, only bio-affinity profiling is applicable at present; i.e., in terms of lectin microarray, an emerging technique in glycotechnology. As described, the technique is straightforward in analyzing both purified glycoproteins (Kuno et al. 2005) and cell lyzates (Ebe et al. 2006), and even intact live cells without disruption (Hsu et al. 2006; Tateno et al. 2007). To perform lectin microarray, glycoproteins are labeled with appropriate dyes, e.g., Cy3 or Cy5, widely used for proteomics. After removal of excess dyes, labeled glycoproteins are subjected to interaction analysis by contacting with immobilized lectins on a glass slide. In fact, there have been an increasing number of reports describing development of lectin microarray since 2005. Among them, however, only evanescent-filed activated fluorescence detection method (Kuno et al. 2005; Uchiyama et al. 2006) enables liquid-phase detection, and requires no washing procedure (for details, see Chapter by Kuno and Hirabayashi in *Experimental Glycoscience: glycobiology*). As a classical method to profile cell surface glycans, is flow-cytometry using fluorescently labeled lectins. Since the developed microarray system has numerous practical merits over flow-cytometry as well as potentially highly useful applications including the use of antibody (e.g., sandwich procedure), the method will greatly contribute to pioneering new fields of glycoscience and glycotechnology.

References

Ebe Y, Kuno A, Uchiyama N, Koseki-Kuno S, Yamada M, Sato T, Narimatsu H, Hirabayashi, J (2006) Application of lectin microarray to crude samples: differential glycan profiling of lec mutants. J Biochem (Tokyo) 139:323–327

Hirabayashi J (2004) Lectin-based structural glycomics: glycoproteomics and glycan profiling. Glyco-conj J 21:35–40

Hsu KL, Pilobello KT, Mahal LK (2006) Analyzing the dynamic bacterial glycome with a lectin microar-ray approach. Nat Chem Biol 2:153–157

Kuno A, Uchiyama N, Koseki-Kuno S, Ebe Y, Takashima S, Yamada M, Hirabayashi J (2005) Evanes-cent-field fluorescence-assisted lectin microarray: a new strategy for glycan profiling. Nat Methods 2:851–856

Nakamura-Tsuruta S, Uchiyama N, Hirabayashi J (2005) High-throughput analysis of lectin–oligosac-charide interactions by automated frontal affinity chromatography. Methods Enzymol 415:311–625

Shimaoka H, Kuramoto H, Furukawa J, Miura Y, Kurogochi M, Kita Y, Hinou H, Shinohara Y, Nishimura S (2007) One-pot solid-phase glycoblotting and probing by transoximization for high-throughput glycomics and glycoproteomics. Chemistry 13:1664–1673

Tateno H, Uchiyama N, Kuno A, Togayachi A, Sato T, Narimatsu H, Hirabayashi J (2007) A novel strategy for mammalian cell surface glycome profiling using lectin microarray. Glycobiology 17:1138–1146

Uchiyama N, Kuno A, Koseki-Kuno S, Ebe Y, Horio K, Yamada M, Hirabayashi J (2006) Development of a lectin microarray based on an evanescent-field fluorescence principle. Methods Enzymol 415:341–351

Monoclonal Antibody as a Clue to Structural Analysis of Bioactive Functional Glycoconjugates

Reiji Kannagi,[1] **Naoko Kimura**[2]

Introduction

Studies using monoclonal antibodies sometimes provide useful information in structural analysis of bioactive functional glycoconjugates. Here, we briefly describe our recent experience on application of anti-carbohydrate antibodies for identification of glycan structures.

Principle

L-selectin ligand expressed on high endothelial venules (HEV) had been assumed to be sialyl Lewis X-like glycans because they are sometimes detectable by anti-sialyl Lewis X antibodies (Sawada et al. 1993). The problem was that only a few anti-sialyl Lewis X could detect the L-selectin ligand on HEV, but others could not. This implied that the L-selectin ligand on HEV, while very similar to conventional sialyl Lewis X, is not exactly the same.

Upon closer examination, the anti-sialyl Lewis X antibodies that detected HEV L-selectin ligands turned out to be reactive to 6-sulfated derivatives of sialyl Lewis X. In contrast, the anti-sialyl Lewis X antibodies that did not detect HEV L-selectin ligands failed to react to the 6-sulfated derivatives. This finding raised the possibility that the L-selectin ligand carried by HEV could be 6-sulfated derivatives of sialyl Lewis X.

Next, we newly generated several antibodies specific to each 6-sulfated derivative of sialyl Lewis X. By using these antibodies, the HEV L-selectin ligand was demonstrated to be sialyl 6-sulfo Lewis X, because only anti-sialyl 6-sulfo Lewis X detected HEV L-selectin ligand and functionally inhibited binding of L-selectin to this in situ ligand (Fig. 1, Mitsuoka et al. 1998).

Since then nearly 10 years have elapsed, and this conclusion was ascertained by reconstitution experiments using co-transfection of genes for fucosyltransferase and GlcNAcβ:6-O-sulfotransferase (Kimura et al. 1999), and also by the generation of mice deficient in GlcNAcβ:6-O-sulfotransferase (Uchimura et al. 2005) during the past decade. Chemical characterization of the ligand structure by mass spectrometry and/or NMR, however, is still just in the primitive stages. This is because the high endothelial venules comprise only a minor population in peripheral lymph nodes, and it is not easy to obtain a large amount of starting material for NMR and even for mass spectrometric analyses. Even if a sufficient number of HEV endothelial cells were to be prepared, sialyl 6-sulfo Lewis X would still be a very minor component of total cellular glycans.

Department of Molecular Pathology, Aichi Cancer Center, Japan
Phone: +81-52-762-6111, Fax: +81-52-764-2973
E-mail: [1]rkannagi@aichi-cc.jp, [2]nkimura@aichi-cc.jp

Fig. 1 Identification of L-selectin ligand in high endothelial venules of human peripheral lymph nodes as sialyl 6-sulfo Lewis X. **a** Structure of sialyl 6-sulfo Lewis X; **b** anti-sialyl 6-sulfo Lewis X specificity of G152 antibody as ascertained by TLC-immunostaining; **c** detection of sialyl 6-sulfo Lewis X in high endothelial venules of human peripheral lymph nodes using G152 antibody

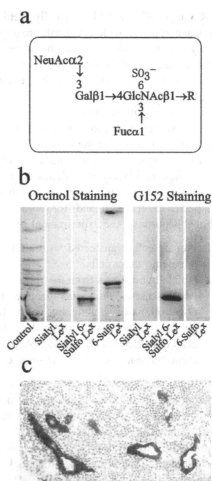

The monoclonal antibody approach is useful for initial structural identification of functional glycans present as a minor component in a small population of cells. A similar example is identification of 6-sulfated glycans in human peripheral blood dendritic precursor cells, or that of cyclic sialic acid residue in a small subset of human lymphocytes (Mitsuoka et al. 1999).

Protocols

Given a sufficient amount of purified carbohydrate antigen on hand, and the wish to obtain a specific monoclonal antibody, it is not difficult to generate an antibody directed to a specific carbohydrate determinant. The classical immunization protocols employing antigen coated on acid-treated *Salmonella minnesota* R595 strain (Kannagi 2000), or antigenic liposomes containing monophosphoryl lipid A (MPL) and trehalose (TDM)

(Kannagi 2002) would eventually yield successful results. However, biologically functional carbohydrate determinants are usually very minor components, and it is not always easy to obtain sufficient amounts of purified materials for immunization. Moreover, the exact structure of the functional component on the cell surface is not always clear at the early stages of investigation. The generation of antibodies directed to minor functional carbohydrate determinants, which can serve as a clue to their structural identification, is a matter of trial and error.

As we had generated a series of monoclonal antibodies directed to $\alpha2 \rightarrow 3$ sialyl 6-sulfo determinants for the study of L-selectin ligands (Mitsuoka et al. 1998), recently we started to systematically generate antibodies directed to 6-sulfo determinants having sialic acid residue(s) other than $\alpha2 \rightarrow 3$ linkage, possibly $\alpha2 \rightarrow 6$ sialyl 6-sulfo determinants, $\alpha2 \rightarrow 8\alpha2 \rightarrow 3$ disialyl 6-sulfo determinants, etc. These structures were expected to serve as ligands for some carbohydrate-recognition molecules such as Siglecs. For this purpose, the experimental protocols applied were as follows.

1. Prepare cells transfected with a GlcNAcβ: 6-O-sulfotransferase gene and immunize mice with these cells.
2. Obtain hybridomas from the spleen of immunized mice and screen for the hybridoma which secretes antibodies reactive only to the cells transfected with a GlcNAcβ: 6-O-sulfotransferase gene, but not to the non-transfected parental cells.
3. Perform further screening for the hybridoma secreting antibodies reactive only to the cells transfected with the GlcNAcβ: 6-O-sulfotransferase gene after sialidase S treatment, but not reactive to the same cells treated with sialidase A.

Since sialidase S is specific to the $\alpha2 \rightarrow 3$ sialic acid linkage, and sialidase A cleaves essentially all sialic acid linkages including $\alpha2 \rightarrow 3$, $\alpha2 \rightarrow 6$, $\alpha2 \rightarrow 8$ and $\alpha2 \rightarrow 9$, this procedure must yield a set of hybridomas secreting antibodies directed to sialyl 6-sulfo determinants having other than the $\alpha2 \rightarrow 3$ linkage. All of the obtained hybridomas were stocked for future use. Next, we decided to first look for antibodies directed to $\alpha2 \rightarrow 6$ sialyl 6-sulfo determinants in the stock, because such determinants are expected to be a preferred ligand for Siglec-2 (Blixt et al. 2004).

4. The cells transfected with the GlcNAcβ: 6-O-sulfotransferase gene were treated with sialidase A, and reacted with an $\alpha2 \rightarrow 6$ sialyltransferase together with CMP-NeuAc. The antibodies secreted by the hybridoma obtained in the above procedures were then further screened for reactivity to these cells. Negative controls for this screening included the sialidase-treated cells reacted with an $\alpha2 \rightarrow 3$ sialyltransferase and CMP-NeuAc.

Several hybridomas screened this way were regarded to secrete antibodies directed to $\alpha2 \rightarrow 6$ sialyl 6-sulfo determinants. All of them were effective in inhibiting the binding of siglec-2. Later, we obtained a small amount of pure material having the terminal structure NeuAc$\alpha2 \rightarrow 6$Gal$\beta1 \rightarrow 4$GlcNAc(6-sulfate)$\beta1 \rightarrow 3$Galβ-R, and antibodies secreted by some of those hybridomas were shown to react to this structure, and the specificity of such antibodies was well established (Kimura et al. 2007). The enigmatic, potentially interesting finding is that antibodies from some other hybridomas were not reactive to this synthetic material, but specifically reacted to the cells treated with $\alpha2 \rightarrow 6$ sialyltransferase, and were effective in inhibiting the siglec-2 binding. Perhaps some difference in the reducing end of the determinant may be involved in this phenomenon.

References

Blixt O, Head S, Mondala T, Scanlan C, Huflejt ME, Alvarez R, Bryan MC, Fazio F, Calarese D, Stevens J, Razi N, Stevens DJ, Skehel JJ, van Die I, Burton DR, Wilson IA, Cummings R, Bovin N, Wong CH, Paulson JC (2004) Printed covalent glycan array for ligand profiling of diverse glycan binding proteins. Proc Natl Acad Sci USA 101:17033–17038

Kannagi R (2000) Monoclonal anti-glycosphingolipid antibodies. Methods Enzymol 312:160–179

Kannagi R (2002) Use of liposomes containing carbohydrates for production of monoclonal antibodies. Methods Mol Biol 199:203–218

Kimura N, Mitsuoka C, Kanamori A, Hiraiwa N, Uchimura K, Muramatsu T, Tamatani T, Kansas GS, Kannagi R (1999) Reconstitution of functional L-selectin ligands on a cultured human endothelial cell line by co-transfection of α1 → 3 fucosyltransferase VII and newly cloned GlcNAcβ: 6-sulfotransferase cDNA. Proc Natl Acad Sci USA 96:4530–4535

Kimura N, Ohmori K, Miyazaki K, Izawa M, Matsuzaki Y, Yasuda Y, Takematsu H, Kozutsumi Y, Moriyama A, Kannagi R (2007) Human B-lymphocytes express alpha 2-6 sialylated 6-sulfo-N-acetyllactosamine serving as a preferred ligand for CD22/siglec-2. J Biol Chem 282:32200–32207

Mitsuoka C, Sawada-Kasugai M, Ando-Furui K, Izawa M, Nakanishi H, Nakamura S, Ishida H, Kiso M, Kannagi R (1998) Identification of a major carbohydrate capping group of the L-selectin ligand on high endothelial venules in human lymph nodes as 6-sulfo sialyl Lewis x. J Biol Chem 273:11225–11233

Mitsuoka C, Ohmori K, Kimura N, Kanamori A, Komba S, Ishida H, Kiso M, Kannagi R (1999) Regulation of selectin binding activity by cyclization of sialic acid moiety of carbohydrate ligands on human leukocytes. Proc Natl Acad Sci USA 96:1597–1602

Sawada M, Takada A, Ohwaki I, Takahashi N, Tateno H, Sakamoto J, Kannagi R (1993) Specific expression of a complex sialyl Lewis X antigen on high endothelial venules of human lymph nodes: possible candidate for L-selectin ligand. Biochem Biophys Res Commun 193:337–347

Uchimura K, Gauguet JM, Singer MS, Tsay D, Kannagi R, Muramatsu T, Von Andrian UH, Rosen SD (2005) A major class of L-selectin ligands is eliminated in mice deficient in two sulfotransferases expressed in high endothelial venules. Nat Immunol 6:1105–1113

Microsequencing of Functional Chondroitin Sulfate Oligosaccharides

Kazuyuki Sugahara, Shuhei Yamada

Introduction

Accumulating evidence implies that chondroitin sulfate (CS)/dermatan sulfate (DS) has a wide range of biological activities, which result from the ability to interact with various proteins, such as growth factors, cytokines, and neurotrophic factors, thereby causing their activation, inactivation, or stabilization. CS/DS chains are linear polysaccharides composed of the various disaccharide units shown in Fig. 1 (Nandini and Sugahara 2006). Functional domains with distinct sequences, which are recognized by specific protein ligands, are formed by combinations of these disaccharide units. Here, methods for the microsequencing of CS oligosaccharides with specific biological activities are described.

Principle

Isolation of Oligosaccharides from Functional Domains of CS Chains

To elucidate the structure of functional domains within CS/DS chains, oligosaccharides that bind to the corresponding protein ligands are prepared. Most bioactive oligosaccharides have been prepared by partial fragmentation of CS/DS chains through enzymatic digestion and subsequent separation by affinity chromatography. Partial depolymerization of CS/DS chains is conducted by digestion with CS/DS-degrading enzymes including bacterial chondroitinases, hyaluronidases, and testicular hyaluronidase. Since testicular hyaluronidase is an endohexosaminidase having a broad substrate specificity, it is often used to prepare CS oligosaccharides. Partial digestion with testicular hyaluronidase generates a series of even-numbered CS oligosaccharides bearing a D-glucuronic acid residue at the nonreducing end. The digest is efficiently size-fractionated by gel filtration on a column of Bio-Gel P-10, being monitored by measuring absorbance at 210 nm. Size-defined oligosaccharide fractions are separated using an affinity column, onto which a functional protein of interest is immobilized, into bound and unbound oligosaccharides, and then the bound oligosaccharides are further fractionated into individual components by anion-exchange HPLC on an amine-bound silica column for microsequencing. Alternatively, size-defined oligosaccharide fractions are subfractionated by

Laboratory of Proteoglycan Signaling and Therapeutics, Graduate School of Life Science, Hokkaido University, Sapporo 001-0021, Japan
Phone: +81-11-706-9054, Fax: +81-11-706-9056
E-mail: k-sugar@sci.hokudai.ac.jp

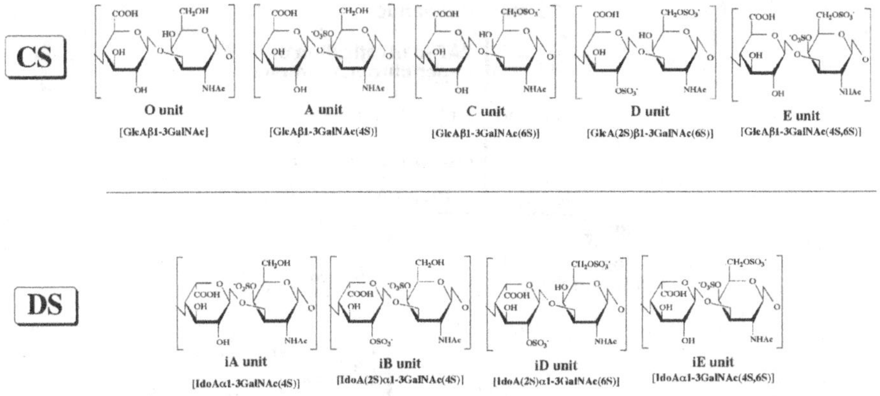

Fig. 1 Structures of the typical *CS/DS* disaccharide units and their terminology. Sulfated disaccharide structures, which are identified in various CS chains as "building units", are shown. 2S, 4S, and 6S represent 2-*O*-sulfate, 4-*O*-sulfate, and 6-*O*-sulfate, respectively

anion-exchange HPLC instead of affinity chromatography, and then each structure of isolated oligosaccharides is determined as described below. Using these structurally defined oligosaccharides, interaction with protein ligands of interest can be examined.

Determination of the Sequence of Isolated Oligosaccharides

For the initial characterization of an isolated CS oligosaccharide, the disaccharide composition is usually analyzed, which reveals basic structural aspects including the proportion of disaccharide units and the total negative charges of the oligosaccharide. The analysis is carried out by anion-exchange HPLC of a chondroitinase ABC digest. In contrast to testicular hyaluronidase, chondroitinases are bacterial CS lyases. In principle, chondroitinase ABC produces the saturated disaccharide unit GlcA-GalNAc as well as the unsaturated disaccharide unit ΔHexA-GalNAc with or without sulfate groups at different positions from the nonreducing as well as the internal and reducing terminal positions, respectively, which are derived from the corresponding disaccharide units depicted in Fig. 1 (Nandini and Sugahara 2006). Since only limited amounts of bioactive oligosaccharides can be isolated, disaccharides are tagged with a fluorophore, 2-aminobenzamide (2AB), at the reducing terminus. The detection limit for a 2AB-derivative is 1–3 pmol. Under the HPLC conditions we have developed, 2AB-derivatives of variously sulfated saturated and unsaturated disaccharides can be separated (Kinoshita and Sugahara 1999). It is noteworthy that a nonreducing terminal disaccharide unit of a parent oligosaccharide can be identified by detection of the 2AB-derivative of the saturated disaccharide, which can be separated from the corresponding 2AB-labeled unsaturated disaccharide counterpart.

The sequential arrangement of disaccharides in a given oligosaccharide can be determined as outlined in Fig. 2 (Deepa et al. 2007a, b), taking advantage of the specificity of chondroitinases. An oligosaccharide is tagged with 2-AB before enzymatic digestion. The 2AB-labeled oligosaccharide is divided into two parts, with each (approx. 30 pmol) being digested with chondroitinase AC-II or ABC, separately, and one-third of each digest is analyzed directly by anion-exchange HPLC. The rest of each digest is further

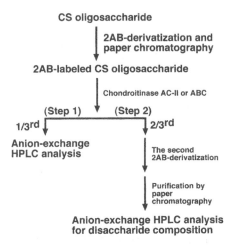

Fig. 2 Strategy for the exo-sequencing of CS oligosaccharides. Individual oligosaccharides fractions (60 pmol) were derivatized with 2AB at the reducing termini. After removal of the excess 2AB reagent by paper chromatography, the 2AB-derivatized oligosaccharides fractions were digested individually with chondroitinase ABC or AC-II. Aliquots (10 pmol each) from these digests were analyzed for 2AB-labeled unsaturated disaccharide or tetrasaccharide products by anion-exchange HPLC to identify the reducing terminal disaccharide unit in each oligosaccharides (*Step 1*), whereas the rest of each digest (20 poml) was further labeled with 2AB, purified by paper chromatography and analyzed by anion-exchange HPLC to identify the nonreducing terminal and the internal disaccharide units (*Step 2*)

labeled with 2AB and analyzed by anion-exchange HPLC to identify the disaccharides to be released from the nonreducing terminal and internal positions.

Chondroitinase AC-II degrades a 2AB-labeled oligosaccharide into disaccharides: one saturated disaccharide derived from the nonreducing terminus, one unsaturated disaccharide tagged with 2AB from the reducing terminus, and unsaturated disaccharides from internal positions. In contrast, chondroitinase ABC cleaves a 2AB-labeled oligosaccharide into one saturated disaccharide from the nonreducing terminal, one 2AB-labeled tetrasaccharide from the reducing terminal, and unsaturated disaccharides from the internal positions. Hence, digestion with chondroitinase AC-II will reveal the reducing terminal disaccharide unit, whereas chondroitinase ABC will give information about the tetrasaccharide from the reducing terminal. A 2AB-labeled tetrasaccharide from the reducing terminus of the parent oligosaccharide generated by the digestion with chondroitinase ABC can be identified by comparing its elution position with those of the authentic 2AB-tetrasacharide standards.

Two-thirds of a digest of a 2AB-labeled oligosaccharide obtained with chondroitinase AC-II or ABC is derivatized again with 2AB to tag the resultant disaccharides derived from the nonreducing terminal and internal positions of the parent oligosaccharide, and the 2AB-labled disaccharides are analyzed by anion-exchange HPLC after purification by paper chromatography to remove the derivatizing reagents (Fig. 2).

This strategy makes it possible to sequence CS tetra-, hexa-, and octasaccharides using less than 100 pmol of the sample within 3 days (Deepa et al. 2007b). Some CS oligosaccharides longer than octasaccharides can be sequenced using this method. Recently, the sequencing of 6 decasaccharides derived from squid cartilage CS-E has been accomplished (Deepa et al. 2007a). For the analysis of larger oligosaccharides or L-iduronic acid-containing DS-type oligosaccharides, other CS lyases such as the chondroitinase ABC protease-free preparation or chondroitinase B, respectively, will be useful. A representative protocol of exo-sequencing of CS octasaccharides is shown as follows.

Exo-sequencing of CS-C Octasaccharides (Deepa et al. 2007b)

Preparation of Oligosaccharide Fractions by Digestion of CS-C with Hyaluronidase

1. Incubate 100 mg of commercial shark cartilage CS-C with 10 mg (approximately 15,000 National formulary units) of sheep testicular hyaluronidase in a total volume of 2.0 ml of 50 mM sodium phosphate buffer, pH 6.0, containing 150 mM NaCl (1 National formulary unit corresponds to the amount of enzyme that hydrolyzes 74 µg of hyaluronate/min) for 36 h.
2. Precipitate proteins by the addition of 0.42 ml of 30% trichroloacetic acid and centrifugation at 2,500 rpm for 10 min.
3. Wash the precipitate with 0.5 ml of 5% trichroloacetic acid.
4. Extract the combined supernatant fluid with 3 ml of diethylether three times to remove trichroloacetic acid.
5. Neutralize the aqueous phase with 1 M Na_2CO_3.
6. Fractionation on a Bio-Gel P-10 column (1.6 × 95 cm), using 1 M NaCl containing 10% ethanol as an eluent.
7. Monitor the eluate by absorbance at 210 nm mainly due to the carbonyl groups.
8. Pool the octasaccharide fraction.
9. Concentrate and desalt the fraction through a Sephadex G-25 (fine) column (1.5 × 46 cm) using distilled water as eluent, being monitored at 210 nm.
10. Quantify by the carbazole assay using GlcUA as a standard.

Subfractionation of the Octasaccharide Fraction by Anion-exchange HPLC

1. Subfractionate the octasaccharide fraction by anion-exchange HPLC on an amine-bound silica PA-03 column (4.6 × 250 mm; YMC Co., Kyoto, Japan) using a linear NaH_2PO_4 gradient from 0.2 to 1 M over a 90-min period at a flow rate of 1 ml/min at room temperature.
2. Monitor the eluate at 210 nm.
3. Collect the separated peaks and desalt individually through a Sephadex G-25 (fine) column (1.5 × 46 cm) using distilled water as eluent.
4. Quantify by the carbazole assay using GlcUA as a standard.

Enzymatic Digestion of Octasaccharides for Determination of Disaccharide Composition

1. Incubate each octasaccharide (0.1 nmol) with 5 mIU of chondroitinase ABC in a total volume of 30 µl in 50 mM Tris–HCl buffer, pH 8.0, containing 60 mM sodium acetate at 37°C for 20 min.
2. Terminate the reaction by boiling for 1 min.
3. Lyophilize the sample in a microcentrifuge tube.

4. Add 5 µl of a 2-aminobenzamide (2AB) derivatization reagent mixture (0.35 M 2AB/1.0 M NaCNBH₄/30% (v/v) acetic acid in dimethyl sulfoxide) to the sample.

5. Incubate the mixture at 65°C for 2 h.

6. Purify the derivatized disaccharides by paper chromatography using Whatman 3 mm paper in a solvent system of butanol:ethanol:water (4:1:1 v/v/v).

7. Analyze the derivatized disaccharides by HPLC on an amine-bound silica PA-03 column using a linear gradient of NaH₂PO₄ from 16 to 800 mM over a 60-min period at a flow rate of 1 ml/min.

8. Identify and quantify the eluted peaks by comparing the elution positions of authentic standards.

Enzymatic Digestion of Octasaccharides for Determination of the Sequence

1. Lyophilize the sample (100 pmol) in a microcentrifuge tube.

2. Add 5 µl of a 2AB derivatization reagent mixture to the dried sample.

3. Incubate the mixture at 65°C for 2 h.

4. Purify the derivatized octasaccharides by paper chromatography using Whatman 3 mm paper in a solvent system of butanol:ethanol:water (4:1:1 v/v/v).

5. Incubate the 2AB-derivatized octasaccharide (30 pmol) with 5 mIU of chondroitinase ABC in a total volume of 30 µl in 50 mM Tris–HCl buffer, pH 8.0, containing 60 mM sodium acetate at 37°C for 20 min to identify the reducing terminal tetrasaccharide.

6. Incubate the 2AB-derivatized octasaccharide (30 pmol) with 5 mIU of chondroitinase AC-II in a total volume of 30 µl in 50 mM acetate buffer, pH 6.0, at 37°C for 20 min to identify the reducing terminal disaccharide.

7. Analyze one-third of chondroitinases ABC and AC-II digests of the 2AB-derivatized octasaccharide (10 pmol) by HPLC on an amine-bound silica PA-03 column using a linear gradient of NaH₂PO₄ from 16 to 800 mM over a 60-min period at a flow rate of 1 ml/min, and identify and quantify the eluted peaks by comparing the elution positions of authentic standards.

8. Lyophilize the remaining two-thirds of the chondroitinase AC-II digest (20 pmol) in a microcentrifuge tube.

9. Add 5 µl of a 2AB derivatization reagent mixture to the dried sample.

10. Incubate the mixture at 65°C for 2 h.

11. Purify the derivatized disaccharides by paper chromatography using Whatman 3 mm paper in a solvent system of butanol:ethanol:water (4:1:1 v/v/v).

12. Analyze the 2AB-derivatives by HPLC on an amine-bound silica PA-03 column using a linear gradient of NaH₂PO₄ from 16 to 800 mM over a 60-min period at a flow rate of 1 ml/min to determine the disaccharide composition and identify the nonreducing terminal disaccharide of the intact octasaccharide.

13. Identify and quantify the eluted peaks by comparing the elution positions of authentic standards.

Based on the data obtained by HPLC analysis, the octasaccharide sequence can be deduced as described in the text.

Applications

The chondroitinase B-resistant octasaccharide fraction was isolated from CS/DS hybrid chains from embryonic pig brains (E-CS/DS) and subjected to pleiotrophin (PTN)-affinity chromatography, because octasaccharides were the minimal size capable of interacting with PTN at a physiological salt concentration. Five and eight octasaccharides were purified from PTN-bound and PTN-unbound octasaccharide fractions, respectively, and their sequences were determined by the method described above. The results revealed a critical role for oversulfated D and/or iD disaccharides in the low yet significant affinity for PTN, which is required for neuritogenesis (Bao et al. 2005).

The sequences of octa- and decasaccharides from squid cartilage CS-E as well as octasaccharides from shark cartilage CS-C were also determined at low picomole amounts by the same method. These structurally defined octa- and decasaccharide fractions were used to investigate the binding specificity of anti-CS antibodies. The results revealed that multiple unique sequences were recognized by each antibody, which implies that the common conformation, molecular shape, and negative charges shared by the multiple primary sequences in the intact CS chains are important as an epitope for each monoclonal antibody (Deepa et al. 2007a, b; Blanchard et al. 2007).

References

Bao X, Muramatsu T, Sugahara K (2005) Demonstration of the pleiotrophin-binding oligosaccharide sequences isolated from chondroitin sulfate/dermatan sulfate hybrid chains of embryonic pig brains. J Biol Chem 280:35318–35328

Blanchard V, Chevalier F, Imberty A, Leeflang BR, Basappa, Sugahara K, Kamerling JP (2007) Conformational studies on five octasaccharides isolated from chondroitin sulfate using NMR spectroscopy and molecular modeling. Biochemistry 46:1167–1175

Deepa SS, Kalayanamitra K, Ito Y, Kongtawelert P, Fukui S, Yamada S, Mikami T, Sugahara K (2007a) Novel sulfated octa- and decasaccharides from squid cartilage chondroitin sulfate-E: sequencing and their application for determination of the epitope structure of monoclonal antibody MO-225. Biochemistry 46:2453–2465

Deepa SS, Yamada S, Fukui S, Sugahara K (2007b) Structural determination of novel sulfated octasaccharides isolated from chondroitin sulfate of shark cartilage and their application for characterizing monoclonal antibody epitopes. Glycobiology 17:631–645

Kinoshita A, Sugahara K (1999) Microanalysis of glycosaminoglycan-derived oligosaccharides labeled with the fluorophore 2-aminobenzamide by high-performance liquid chromatography: application to disaccharide composition analysis and exo-sequencing of oligosaccharides. Anal Biochem 269:367–378

Nandini CD, Sugahara K (2006) Role of the sulfation pattern of chondroitin sulfate in its biological activities in the binding of growth factors. In: Volpi N (ed) Advances in pharmacology, vol 53. Elsevier, UK, pp 253–279

Structural Characterization of PA-oligosaccharide Isomers Derived from Glycosphingolipids by MALDI-TOF Mass Spectrometry

Minoru Suzuki

Introduction

For the analysis of oligosaccharides derived from glycosphingolipids, a method involving pyridylamino (PA) derivatization, normal- and reversed-phase HPLC with fluorescent detection (Hase 1994) after treatment with endoglucoceramidase (Ito and Yamagata 1986) is widely used. In order to characterize structures of the PA-oligosaccharides, matrix-assisted laser desorption/ionization quadrupole ion trap time-of-flight mass spectrometry (MALDI-QIT-TOF MS), which has the capability to perform MS^n analysis, has been successfully used (Suzuki et al. 2005).

Results

The MALDI-QIT-TOF MS was used to analyze three linkage isomers of PA-oligosaccharides: lacto-N-fucopentaose (LNFP) I, II, and III. In the positive-ion MALDI-QIT-TOF mass spectra of the PA-oligosaccharides, [M + Na] ions were observed at m/z 954, and the mass spectra of these oligosaccharide isomers are the same. In the MS^2 spectra selecting [M + Na] ion at m/z 954 as the precursor ion shown in Fig. 1, m/z 792 was not detected in LNFP I, and the relative intensities of m/z 388 and 696 in LNFP I were different from those of LNFP II and III. The difference in relative intensities of m/z 406 and 550 was observed for LNFP II and III, respectively. However, these relative intensities of the fragment ions were not always reproducible. Figure 2a–c show the MS^3 spectra derived from the second precursor ion at m/z 534 of LNFP I, II, and III, respectively. In the MS^3 mass spectrum of LNFP I, fragment ions at m/z 388 (B3/Y4 + Na), 349 (C2 + Na), and 331 (B2 + Na) were observed. By contrast, the fragment ions at m/z 388(B2/Y3β + Na), 372 (B2/Y3α + Na), and 354 (B2/Z3α + Na) were detected in the MS^3 mass spectrum of LNFP II, and the ions at m/z 388 (B2/Y3β + Na) and 370 (B2/Z3β + Na) were obtained in the MS^3 mass spectrum of LNFP III. The specific fragment ions are m/z 349 (C + Na) for LNFP I, 372 (B2/Y3α + Na) and 354 (B2/Z3α + Na) for LNFP II, and 370 (B2/Z3β + Na) for LNFP III. These results indicate that the structures of these oligosaccharide isomers could be characterized using the MS^3 analysis.

Sphingolipid Expression Laboratory, RIKEN Frontier Research System, 2-1 Hirosawa, Wako, Saitama 351-1098, Japan
Phone: +81-48-467-4404, Fax: +81-48-462-4615
E-mail: msuzuki@riken.jp

Fig. 1 MS^2 spectra of LNFP I (**A**), II (**B**), and III (**C**) derived from precursor ions at *m/z* 954

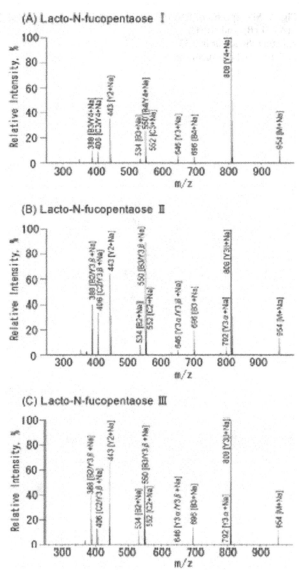

(A) Lacto-N-fucopentaose I

(B) Lacto-N-fucopentaose II

(C) Lacto-N-fucopentaose III

Procedures

1. Prepare glycosphingolipid fraction.
2. Digest glycosphingolipids with endoglucoceramidase.
3. Dry upper layer solution obtained after Folch's partition of the digests.
4. Derivatize with 2-aminopyridine (PA) derivatization kit (Takara, Kyoto, Japan) to PA-oligosaccharides.
5. Purify of each PA-oligosaccharide by preparative-HPLC.
6. Apply 1–2 μl of the solution on a MALDI-TOF MS plate.
7. Dry the solution with a gentle stream of air.
8. Introduce the plate to a MALDI-TOF mass spectrometer (Axima MALDI-QIT-TOF, Shimadzu Corp., Kyoto, Japan).

Fig. 2 MS^3 spectra of LNFP I (**A**), II (**B**), and III (**C**), selecting the ion at m/z 534 as the second precursor ion

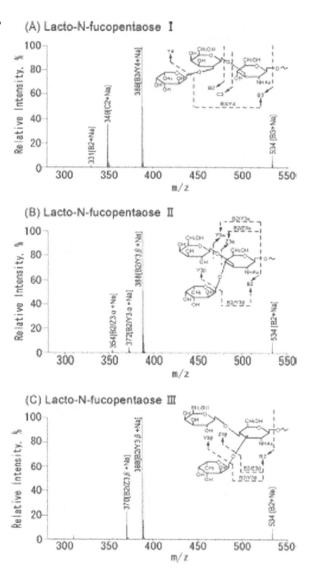

(A) Lacto-N-fucopentaose I

(B) Lacto-N-fucopentaose II

(C) Lacto-N-fucopentaose III

9. Positive ion mode MALDI-TOF MS measurement with a 337 nm pulsed N2 laser.
10. MS/MS measurement selecting Na adduct ion at m/z 954 as the precursor ion.
11. MS/MS/MS measurement selecting an ion at m/z 534 as the precursor ion.

Comment

MS/MS/MS analysis of the isomer of PA-oligosaccharide can provide more clear structural information than that of MS/MS.

References

Hase S (1994) High-performance liquid chromatography of pyridyl-aminated saccharides. Methods Enzymol 230:225–237

Ito M, Yamagata T (1986) A novel glycosphingolipid degrading enzyme cleaves the linkage between the oligosaccharide and ceramide of neutral and acidic glycosphingolipids. J Biol Chem 261: 14278–14282

Suzuki Y, Suzuki M, Ito E, Ishii H, Miseki K, Suzuki A (2005) Convenient and rapid analysis of linkage isomers of fucose-containing oligosaccharides by matrix-assisted laser desorption/ionization quadrupole ion trap time-of-flight mass spectrometry. Glycoconj J 22:427–431

Structural Analysis of Phospho-Glycosphingolipids in Lower Animals

Saki Itonori, Mutsumi Sugita

Introduction

Phosphoric acid is a key compositional element of ATP which provides biochemical energy as well as contributes to the nucleic acid backbone in all organisms. It is also an indispensable element in the phospholipid component of biomembranes. Phospho-glycosphingolipid (PGSL) contains phosphoric acid as a charged group (Fig. 1). Numerous PGSLs have been reported in invertebrates, and it is possible that they are the functional counterpart in invertebrates to the acidic glycosphingolipid, ganglioside in vertebrates. It is thus important to gain a better understanding of PGSL compounds and their structures, and their distribution in the animal kingdom (Hori and Sugita 1993; Itonori and Sugita 2005; Itonori and Sugita 2007).

Distribution

The first reported chemical structure of PGSL was a galactosylceramide containing *N*-methyl aminoethylphosphonic acid (MAEPn) from the marine snail, *Turbo cornutus* (Mollusca, Gastropoda). From the sea hare (Mollusca, Gastropoda), a group of PGSLs has been characterized containing 1–3 mol of aminoethylphosphonic acid (AEPn). Mollusca contain predominantly ceramide aminoethylphosphonic acid as the major phospho-sphingolipid containing the C–P bond. Antarctic krill and the marine crab (Arthropoda, Crustacea) produce PGSLs containing MAEPn or AEPn in which the polar group is attached to the 4- or 6-hydroxy group of GlcCer.

Another characteristic phosphosphingolipid in invertebrates is ceramide phosphoethanolamine, whose known distribution so far is limited to Arthropoda and a group of Mollusca. A series of PGSLs containing phosphoethanolamine (Etn*P*) has been characterized from the fly (Arthropoda, Insects) in which Etn*P* is linked to the 6-hydroxy group of the GlcNAc moiety of Arthro series GSLs. PGSLs containing both uronic acid and Etn*P* or 2 mol of Etn*P* also have been characterized from the fly or *Drosophila melanogaster.*

A series of PGSLs containing choline phosphate (PC) has been characterized from the earthworm (Annelida, Oligochaeta) as zwitterionic GSLs. This discovery has led to the finding of other zwitterionic GSLs in other annelida e.g., the marine annelid, lugworm, and the freshwater leech or nematoda. From the porcine parasitic roundworm, *Ascaris suum* and the free-living nematode, *Caenorhabditis elegans* (Nematoda), a series of PGSLs has been characterized having PC attached to the 6-hydroxy of GlcNAc on their

Department of Chemistry, Faculty of Liberal Arts and Education, Shiga University, 2-5-1 Hiratsu, Otsu, Shiga 520-0862, Japan
Phone: +81-77-537-7728, Fax: +81-77-537-7728
E-mail: itonori@sue.shiga-u.ac.jp

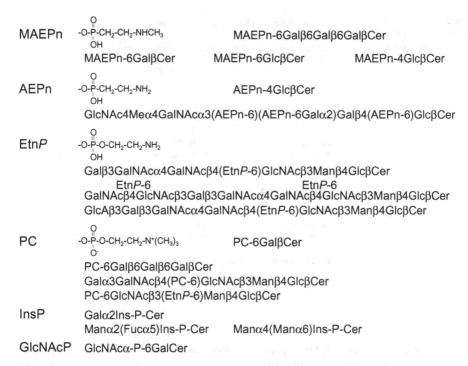

Fig. 1 Characteristic phospho-glycosphingolipids in lower animal

characteristic neutral GSLs; a PGSL has also been reported containing both PC and Etn*P* from the roundworm. It is worth noting that PC-containing PGSLs have been reported to induce the release of proinflammatory monokines.

From the same roundworm, inositol phosphate (InsP)-containing PGSL, a so-called "phytoglycolipid or mycoglycolipid" has been characterized for the first time from Animalia. PGSLs containing InsP have also been characterized from the marine annelid and lugworm (Annelida), in which Fuc and Man are linked to inositol. Thus, the distribution of these PGSLs is not limited to plants and fungi.

A PGSL containing GlcNAc phosphate (GlcNAcP) has been identified as an acidic GSL from the liver fluke *Fasciola hepatica*; this PGSL is a highly antigenic determinant against infected sera.

Sample Preparation (Fig. 2)

1. Soak the tissues in 1.5–2 volumes of acetone for 2 h (removal of the simple lipids and the internal water). Remove the mixture of acetone and water by decantation and repeat these procedure two times.
2. Dry the tissues under atmosphere and homogenize the dried tissues with a metal blender.
3. Extract the tissue powder with 3–5 volumes of chloroform–methanol (C/M, 2:1, v/v).
4. Extract the residue further with C/M = 2:1 and 1:1. Combine the extracts and evaporate by a rotary evaporator.

Fig. 2 Procedure for fractionation of PGSL from invertebrate animal

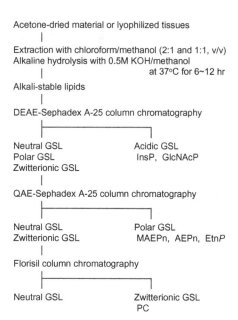

5. Apply the sample on DEAE-Sephadex (acetate form) column chromatography with chloroform–methanol–water (C/M/W, 30:60:8, v/v) and elute the acidic fraction with 0.05, 0.15, and 0.45 M ammonium acetate in methanol.
6. Apply the pass-through fraction from DEAE resin onto QAE-Sephadex (OH⁻ form) column chromatography with C/M/W = 30:60:8 and elute the polar fraction with 0.05, 0.15 and 0.45 M ammonium acetate in methanol.
7. Acetylate the pass-through fraction from QAE resin with pyridine-acetic anhydride (3:2, v/v) for overnight. After removal of solvent, separate acetylated neutral and zwitterionic GSLs by Florisil column chromatography.
8. Isolate the individual GSL components by silicic acid (Iatrobeads) column chromatography and HPLC with C/M/W or propanol–water–ammonia solvent systems.

References

Hori T, Sugita M (1993) Sphingolipids in lower animals. Prog Lipid Res 32:25–45
Itonori S, Sugita M (2005) Diversity of oligosaccharide structures of glycosphingolipids in invertebrates. Trend Glycosci Glycotech 17:15–25
Itonori S, Sugita M (2007) C. Biochemical aspects: glycophylogenetic aspects of lower animals. In: Kamerling JP (ed) Comprehensive glycoscience. Elsevier, UK, pp 253–284

Structural Analysis of Polysialic Acid

Chihiro Sato[1], Ken Kitajima[2]

Introduction

Polysialic acid is the polymerized structure of sialic acid, and its degree of polymerization (DP) varies from 8 to 200. Polysialic acid was first identified in the glycocalyx of neuroinvasive bacteria. Five glycoproteins, fish egg polysialoglycoprotein (PSGP), neural cell adhesion molecule (NCAM), a voltage-gated sodium channel of eel, CD36 in human milk, and neuropilin-2 have been identified as polysialic acid-containing glycoproteins until recently (Sato 2004). By virtue of their net negative charge at physiological pH, they serve as regulators of ligand–receptor and cell–cell interactions via anti-adhesive effect (Bonfanti 2006). Polysialylation occurs in some tumors and is involved in metastasis. Since polysialic acid is directly involved in the biologic function, it is important to analyze the precise structure of polysialic acid. When samples containing di- oligo- and polysialic acid structures at 10–100 µg are analyzed, conventional methods including methylation analysis, NMR, and mild acid hydrolysis-TLC (Sato et al. 1993) can be applied. However, the amount of these types of glycoproteins is often too small to be analyzed by the conventional methods. The fact that di/oligo/polysialic acid-modification of glycoproteins have not been reported so often suggests the rare occurrence of polysialic acid in an organism. As highly sensitive chemical and immunochemical methods to analyze minute amounts of di-, oligo-, and polysialic acid have been developed (Sato et al. 1998, 1999; Sato 2004), studies on di/oligo/polysialic acid have gradually emerged. As chemical detection methods are described in the Chapter by C. Sato this volume, immunochemical methods to detect and determine the polysialic acid structures with specific antibodies and enzymes are introduced as powerful tools in this chapter.

Procedures

1. Apply the glycoprotein samples (denatured at 60°C for 20 min) for SDS-PAGE.
2. Electrophorese on the gel.
3. Transfer proteins from the gel to a PVDF membrane.
4. Wash the PVDF membrane with phosphate buffered saline (PBS) for 10 min.
5. Wash the PVDF membrane with the buffer to be used for the sialidase treatment for 10 min.

Laboratory of Animal Cell Function, Bioscience and Biotechnology Center, Nagoya University, Nagoya 464-8601, Japan
Phone: +81-52-789-4295, Fax: +81-52-789-5228
E-mail: [1]chi@agr.nagoya-u.ac.jp, [2]kitajima@agr.nagoya-u.ac.jp

6. Incubate the PVDF membrane with and without sialidase reagent[1] at 37°C for 20 h.
7. Wash the PVDF membrane with PBS containing 0.05% Tween 20 (PBST).
8. Block the PVDF membrane with 1% BSA in PBST at 37°C for 1 h.
9. Incubate the PVDF membrane with the anti-oligo/polysialic acid antibody (dilute 1–10 μg/ml with 1% BSA in PBST) at 4°C overnight.
10. Wash the PVDF membrane twice with PBST for 10 min.
11. Incubate with enzyme-linked secondary antibodies (dilute with 1% BSA in PBST) at 37°C for 1 h.
12. Wash the PVDF membrane three times with PBST for 10 min.
13. Visualize the glycoproteins with appropriate methods as described elsewhere.

Comments

It is most important to understand precise specificity of the antibodies recognition, if the antibody and enzymes are used for structural determination and functional analysis of carbohydrates. The antibody's specificity toward degree of polymerization is very important before use as immunochemical probes. However, few antibodies are characterized for their immunospecificity. We thus determined the immunospecificity of so-called anti-polysialic acid antibodies and newly developed antibodies by an ELISA-based method using a series of phosphatidylethanolamine-conjugated oligo/polysialic acid with the defined sialic acid composition and DP as the test antigens (Sato et al. 1995). Using this method, the precise immunospecificity of various anti-di/oligo/polysialic acid antibodies has been successfully classified into three groups based on the DP and non-reducing terminal end for antibody recognition (Table 1). Group I antibodies are the "anti-polysialic acid antibodies" that recognize chains of $\alpha2 \rightarrow 8$-linked sialic acid with DP 8 or greater sialic acid residues, including fully extended polysialic acid chains. These antibodies recognize the helical conformation formed by sialic acid residues within the internal region of the polysialic acid chains. The non-reducing terminal residues are not involved in antigen recognition. Group II antibodies, designated "anti-oligo + polysialic acid antibodies," recognize both oligosialic acid with DP 2–7 and polysialic acid chains. These antibodies recognize the distal portion of oligo/polysialic acid chains, including the non-reducing termini. Group III antibodies, designated "anti-oligosialic acid antibodies," recognize oligosialic acid with DP 2–4, but do not bind with polysialic acid. These antibodies appear to recognize specific conformations of di/oligosialic acid with DP 2–4. Group III antibodies can be further classified into two subgroups. Actually, in combinational use of these antibodies, we can estimate the degree of polymerization of the glycoconjugates.

Since an antibody, especially an anti-carbohydrate antibody, sometimes binds to some glycoproteins non-specifically (Yasukawa 2006), it is important to make sure the degradation by specific enzymes to confirm the presence of polysialic acid chain. Endo-sialidase can serve as a specific molecular probe to detect and selectively modify $\alpha2 \rightarrow 8$-linked polysialic acid chains. The soluble enzyme derived from bacteriophage K1F, designated Endo-N catalyzes the depolymerization of polysialic acid chains:

---

[1] Sialidase reagent: exo-sialidase (0.5 units/ml in 50 mM sodium acetate buffer, pH 5.0, for example, *Clostridium perfringens* sialidase), endo-sialidase (1 milliunits/ml in 50 mM Tris–HCl, pH 7.5, for example, sialidase from Bacteriophage K1F)

Table 1 Classification of anti-oligo/polysialic acid antibodies

Name of antibody	Animal origin and immunoglobulin type	Procedure for immunization	Immunogen	Sia in oligo/polysialic acid recognized	Specificity on DP
<Group I> anti-polysialic acid antibody					
H.46	ho, poly, IgM	i.v.	*Neisseria meningitidis GpB*	Neu5Ac	DP ≥ 8
735	mo, mono, IgG2a	i.p.	*Neisseria meningitidis GpB*	Neu5Ac	DP ≥ 11
<Group II> anti-oligo + polysialic acid antibody					
12E3	mo, mono, IgM	i.p.	Embryonic rat forebrain	Neu5Ac	DP ≥ 5
5A5	mo, mono, IgM	i.p.	Membrane from embryonic rat spinal cord	Neu5Ac	DP ≥ 3
2-2B	mo, mono, IgM	i.v. and i.p.	*Neisseria meningitidis GpB*	Neu5Ac	DP ≥ 4
OL.28	mo, mono, IgM		Oligodendrocyte from new born rat	Neu5Ac	DP ≥ 4
2-4B	mo, mono, IgM	i.v.	Oligo/polyNeu5Gc-PE	Neu5Gc	DP ≥ 2
Kdn8kdn	mo, mono, IgM	i.p.	KDN-gp	KDN	DP ≥ 2
<Group III> anti-oligoSia antibody					
S2-566	mo, mono, IgM	i.p.	Human GD3	Neu5Ac	DP = 2[a]
1E6	Mo, mono, IgM	i.v.	(Neu5Ac)$_2$-bearing artificial glycopolymer	Neu5Ac	DP = 2
AC1	mo, mono, IgG3	i.v.	(Neu5Gc)GD1c	Neu5Gc	DP = 2-4

[a] *ho* horse, *mo* mouse, *poly* polyclonal, *mono* monoclonal

[a] Neu5Acα2 → 8Neu5Acα2 → 3Gal. Gal residue is required

Table 2 Reactivity of di-, oligo-, and polysialic acid chains toward biochemical probes

Biochemical probes	diSia (α2,8) DP = 2	oligoSia (α2,8) DP = 3 – 7	polysialic acid (α2,8) DP ≥ 8	polysialic acid (α2,9) DP ≥ 8
Group I antibody	–	–	+	–*
Group II antibody	–	+	+	–
Group III antibody	+	–**	–	–
Endo-sialidase (Endo-N) (DP ≥ 5)	–	–	+	–
Endo-sialidase (DP ≥ 3)***	–	+	+	–
Exo-sialidase (α2,3-specific)	–	–	–	–
Exo-sialidase (α2,3,6-specific)	–	–	–	–
Exo-sialidase (α2,3,6,8-specific)	+	+	+	–
Exo-sialidase (α2,3,6,8,9-specific)	+	+	+	+

+reactive or sensitive; – unreactive or insensitive
*Anti-α2,9-linked polysialic acid antibodies are developed; however the immunospecificity was not yet determined
**AC1 can recognize DP3 and 4 Neu5Gc oligomer as well
***Miyake et al. (1997)

$$(\rightarrow 8\text{Neu5Acyl}\alpha 2\rightarrow)_n\text{-}X(n \geq 5) \rightarrow (\rightarrow 8\text{Neu5Acyl}\alpha 2\rightarrow)_{2-4} + (\rightarrow 8\text{Neu5Acyl}\alpha 2\rightarrow)_2\text{-}X$$

Since pentaNeu5Acyl structure is the minimum size required for Endo-N recognition (Hallenbeck et al. 1987), we can discriminate di- or oligoSia (DP 3–5) from polysialic acid (DP ≥ 8), based on effects of Endo-N digestion on immunoreactivity toward Group II antibodies (Table 1). There are other endo-N-acylneuraminidases whose specificity is DP ≥ 3 from bacteriophage, and these probes can be also used for the DP determination (Table 2). The Endo-N insensitive di- and oligoSia structures should be cleaved by linkage-specific sialidases (α2,3-specific sialidase, α2,6-specific sialidase, α2,3,6,8-specific sialidase, α2,3,6,8,9-specific sialidases). Thus, we can confirm the chain length of di/oligo/polysialyl structures more precisely by the combinational application of the specific antibodies and enzymes (see Table 2) even if the amount of the sample is too small for the chemical methods. Furthermore, in conjunction with chemical and immunochemical methods, the occurrence of di- and oligo- and polysialic acid epitopes on glycoproteins and glycolipids can be detected in amounts as small as 10 fmol of diSia. Now we can determine the precise DP of polysialic acid on the biological active materials. The production of the highly functional materials using sialyloligo/polymers will benefit our lives in near future. Biological meanings of the diversity of the polysialic acid structure (degree of polymerization, component of Sia, and linkage) will also be revealed.

References

Bonfanti L (2006) PSA-NCAM in mammalian structural plasticity and neurogensis. Prog Neurobiol 80:129–164
Hallenbeck PC, Vimr ER, Yu F, Bassler B, Troy F A (1987) Related articles, links purification and properties of a bacteriophage-induced endo-N-acetylneuraminidase specific for poly-alpha-2,8-sialosyl carbohydrate units. J Biol Chem 262:3553–3561
Miyake K, Muraki T, Hattori K, Machida Y, Watanabe M, Kawase M, Yoshida Y, Iijima, S (1997) J Ferm Bioeng 84:90–93
Sato C (2004) Chain length diversity of sialic acids and its biological significance. Trends Glycosci Glycotech 16:331–344

Sato C, Kitajima K, Tazawa I, Inoue Y, Inoue S, Troy FA 2nd (1993) Structural diversity in the alpha 2 → 8-linked polysialic acid chains in salmonid fish egg glycoproteins. Occurrence of poly(Neu5Ac), poly(Neu5Gc), poly(Neu5Ac, Neu5Gc), poly(KDN), and their partially acetylated forms. J Biol Chem 268:23675–23684

Sato C, Kitajima K, Inoue S, Seki T, Troy FA 2nd, Inoue Y (1995) Characterization of the antigenic specificity of four different anti-(alpha 2 → 8-linked polysialic acid) antibodies using lipid-conjugated oligo/polysialic acids. J Biol Chem 270:18923–18928

Sato C, Inoue S, Matsuda T, Kitajima K (1998) Development of a highly sensitive chemical method for detecting α2 → 8-linked oligo/polysialic acid residues in glycoproteins blotted on the membrane. Anal Biochem 261:191–197

Sato C, Inoue S, Matsuda T, Kitajima K (1999) Fluorescent-assisted detection of oligosialyl units in glycoconjugates. Anal Biochem 266:102–109

Yasukawa Z, Sato C, Kitajima K (2007) Identification of an inflammation-inducible serum protein recognized by anti-disialic acid antibodies as carbonic anhydrase II. J Biochem 141:429–441

Milk Oligosaccharides: Structural Characterization and Future Aspects

Tadasu Urashima

Introduction

Milk oligosaccharides are a variety of free saccharides which are found in milk/colostrum along with lactose, and have a lactose unit at their reducing ends. Human milk contains about 7% of carbohydrate of which around 20% consists of more than 100 milk oligosaccharides. Detailed structural studies of human milk oligosaccharides were started by Richard Kuhn during the 1950s, followed by those of Montreuil et al., Kobata et al., Egge et al., Strecker et al., Gronberg et al. and Kitagawa et al., etc. At least 93 human milk oligosaccharides have been characterized up to date (Urashima et al. 1997). Non-human milk oligosaccharides have been studied by Kuhn et al., Messer et al., and Urashima et al., etc., and more than 100 oligosaccharides were characterized in milk or colostrum of brown capuchin, buffalo, horse, cow, goat, sheep, Ezo brown bear, Japanese black bear, polar bear, whitenosed-coati, elephant, rat, dog, beluga, Minke whale, giant panda, crabeater seal, hooded seal, bearded seal, harbor seal, mink, bottlenose dolphin, echidna, platypus, and tammar wallaby (Urashima et al. 2001).

Concept

Prior to characterization of their chemical structures, the oligosaccharides have to be separated from each other and purified. Previously, the separation was usually performed using paper or thin layer chromatography, but more recently normal or reverse phase high performance liquid chromatography techniques (Nakamura et al. 2001) or high pH anion exchange chromatography with pulsed amperometric detection (HPEAC-PAD) (Coppa et al. 1999) have been employed for their separation. For structural characterization, simple as well as small-scale methods needed to be developed. Many milk oligosaccharides contain the units of ABH antigens, α-Gal, Lewis a, b, x or y, and/or α (2-3)/(2-6) sialyl residues linked to core structures such as lacto-N-tetraose, lacto-N-neotetraose, lacto-N-hexaose, lacto-N-neohexaose, para lacto-N-hexaose, lacto-N-octaose, etc. As some regularities among milk oligosaccharide structures were found, analyses of reporter groups as well as anomer resonances in their 1H-NMR spectra, which had been developed by van Halbeek and Vliegenthart for the oligosaccharides of N-glycans, should be powerful methods along with matrix-assisted laser desorption/ionization time-of-flight mass spectrometry (MALDI-TOFMS) or fast atom bombardment mass spectrometry (FAB-MS) (Urashima et al. 2000). Milk oligosaccharides can also be characterized by 1H-13C correlated spectroscopy (1H-13C COSY), heteronuclear single quantum coherence experiment (HSQC), heteronuclear multiple bond correlation experiments (HMBC),

Graduate School of Food Hygiene, Obihiro University of Agriculture and Veterinary Medicine, Nishi 2sen 11banchi, Inada-cho, Obihiro, Hokkaido 080-8555, Japan
Phone: +81-155-49-5566, Fax: +81-155-49-5577
E-mail: urashima@obihiro.ac.jp

1H–1H homonuclear Hartmann–Hahn experiments (1H–1H HOHAHA), etc. It can be expected that, in the near future, affinity chromatography methods using lectins or mono-clonal antibodies will be applied on a very small scale for the structural characterization of glycoconjugates, including milk oligosaccharides. Much basic kinetic data relating to various lectins and known authentic milk oligosaccharides whose structure is likely to be obtained. In this respect, the approaches by Hirabayashi et al. (2002) and Nakajima et al. (2006) are notable.

The study of milk oligosaccharides had been started for the purpose of finding the bifidobacterium growth-stimulating factor in human milk. As it is known that milk oli-gosaccharides have a prebiotic effect, a few companies have begun to manufacture milk oligosaccharide-like saccharides, such as galacto oligosaccharides.

It is to be noted that the structures of milk oligosaccharides are similar to the carbo-hydrate moieties of glycoconjugates on the surface of epithelial cells. As these glyco-conjugates are receptors for pathogenic bacteria, viruses, and protozoa and for their attachment to the host cells, it is considered that milk oligosaccharides are receptor ana-logues that inhibit the attachment. It can be expected that authentic milk oligosaccharides will be available for the determination of the receptor structures of many pathogenic microorganisms. For example, experiments using poly-galactosylated milk oligosaccha-rides, which had been isolated from tammar wallaby milk, showed that a pathogenic protozoa, *Laishmania major*, has affinity to galectin-3 or galectin-9 (Pelletier et al. 2003). There is a strategic demand for the isolation of many oligosaccharides from human as well as non-human milk that will contribute to library collections that can be used to determine the receptor structures of various human and animal pathogens. Following determination of the receptor structure of a pathogen, an oligosaccharide, whose structure is similar to that of the receptor unit, may be utilized for the prevention of infection, or as therapy. It can be expected that, in future, milk oligosaccharides isolated from milk/colostrum of farm animals, or prepared by recombinant or transgene techniques, will be utilized for the above purpose.

Characterization of Milk Oligosaccharides

The carbohydrate fractions are separated from milk or colostrum by extraction with chloroform/methanol, and the oligosaccharides are purified by gel filtration of the carbohydrate as well as by ion exchange and high performance liquid chromatography (HPLC). Each oligosaccharide is characterized by ^{1}H, ^{13}C, ^{1}H-^{13}C HSQC NMR spectra and MALDI-TOFMS mass spectrometry. The following experimental protocol is based on the characterization of a novel sialyl tetrasaccharide separated from the colostrum of a bottlenose dolphin (Uemura et al. 2005).

Purification of Oligosaccharides

1. Extract the carbohydrate fraction from 15 ml colostrum with four volumes of chloroform/methanol (2:1, v/v).

2. Separate oligosaccharides from the carbohydrate fraction by passage through a Bio Gel P-2 (<45 μm) column (2.6 × 100 cm^2) with distilled water as the eluate, at a flow rate of 15 ml/h. Analyze aliquots (0.5 ml) of each fraction for hexose with the phenol–H_2SO_4 method and for sialic acid with the periodate–resorcinol method. Pool each frac-tion and freeze dry.

Fig. 1 ¹H-¹³C HSQC correlation spectrum of Neu5Ac(α2-3)[GalNAc (β1-4)]Gal(β1-4)Glc separated from colostrum of a bottlenose dolphin. The appropriate part of the 1D ¹H-NMR spectrum, with proton assignments, is shown at the *top* of each plot. Correlated carbons are indicated next to the relevant contours. **a** Part of the anomeric proton spectrum. **b** Part of the middle magnetic field of the 1D ¹H-NMR spectrum

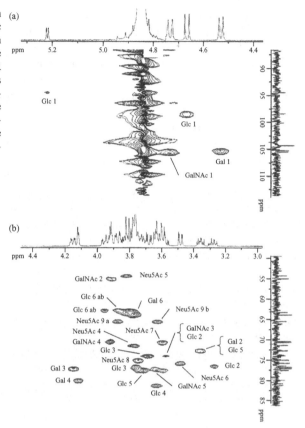

3. Check the saccharides in each fraction by thin-layer chromatography with acetone/2-propanol/0.1 M lactic acid (2/2/1, v/v) as a developing solvent.

4. Separate neutral, monosialyl or disialyl oligosaccharide fractions, which eluate sooner after the void volume during the aforementioned gel chromatography, by anion exchange chromatography with a DEAE–Sephadex A-50 column (1.5 × 35 cm²) using 50 mM Tris hydroxyaminomethane–HCl buffer (pH 8.7). Analyze aliquots of each fraction as above. Pool these fractions and freeze dry. Remove the salt from the pooled fractions by passage through a Bio Gel P-2 column as above.

5. Separate each sialyl oligosaccharide from the above fractions by HPLC using a TSK gel Amido-80 column (4.6 × 250 mm², pore size 80 Å, particle size 5 μm) and a mobile phase of 50 and 80% (v/v) acetonitrile in 15 mM potassium phosphate buffer (pH 5.2) using a linear gradient of acetonitrile from 80 to 50% at 40°C at a flow rate of 1 ml/min. Detect the eluted materials by measuring the absorbance at 195 nm. Remove the salt from each peak component by passage through a Bio Gel P-2 column (1.5 × 35 mm²).

6. Separate each neutral oligosaccharide by HPLC using a TSK gel Amido-80 column and a mobile phase of 50 and 80% (v/v) acetonitrile in ion exchange water using a linear gradient of acetonitrile from 80 to 50% at 40°C at a flow rate of 1 ml/min. Detect the eluted materials with the evaporative light scattering detector at 50°C, gain 8 and 3.5 bar.

Table 1 ^{13}C-NMR chemical shifts of Tt1-4 (A) from bottlenose dolphin colostrum and 3′-SL (B)

Reporter	Chemical shifts (ppm)									
	α-Glc		β-Glc		β-Gal		Neu5Ac		GalNAc	
group	A	B	A	B	A	B	A	B	A	GalNAc
C-1	94.44	94.52	98.39	98.47	105.20	105.29	176.76	176.30	105.41	98.08
C-2	73.73	73.83	76.37	76.49	72.67	72.07	100.11	102.31	54.97	56.32
C-3	73.86	74.09	76.99	77.03	76.65	78.15	39.58	42.23	73.66	73.80
C-4	81.08	80.94	80.70	80.79	79.82	70.17	71.34	70.96	70.40	70.51
C-5	72.70	72.78	77.41	77.50	76.65	77.84	54.23	54.35	77.36	77.86
C-6	63.22	62.59	63.22	62.73	63.81	63.72	75.70	75.60	63.22	63.67
C-7							70.62	70.77		
C-8							74.93	74.35		
C-9							65.47	65.30		
CH_3							24.69	24.71	25.24	24.88
CO							177.65	177.68	177.49	177.65

Characterization of Oligosaccharides

1. Determine the molecular weight of the component with the aid of MALDI-TOFMS.
2. Characterize each oligosaccharide structure from its ^{1}H, ^{13}C- and/or ^{1}H-^{13}C HSQC spectra.

Comments

The ^{1}H-NMR signals of the reporter groups are assigned by comparison with the chemical shifts data published previously for relevant oligosaccharides separated from the milk or colostrum derived from either humans or other species. Each ^{13}C signal is assigned by the correlation of the signal with the ^{1}H signal of the reporter group and by comparison with the chemical shifts data of relevant saccharides. From the assignment of each ^{13}C signal and the ^{1}H signals of the reporter group, the oligosaccharide structures can be characterized. As an example, the ^{1}H-^{13}C HSQC correlation spectrum and the assignment of Neu5Ac(α2-3)[GalNAc(β1-4)]Gal(β1-4)Glc, which was separated from bottlenose dolphin colostrum, are shown in Fig. 1 and Table 1.

References

Coppa GV, Pierani P, Zampini L, Carloni I, Carlucci A, Catacci C, Gabrielli O (1999) Oligosaccharides in human milk during different phases of lactation. Acta Paediatr Suppl 430:89–94

Hirabayashi J, Hashidate T, Arata Y, Nishi N, Nakamura T, Hirashima M, Urashima T, Oka T, Futai M, Muller WEG, Yagi F, Kasai K (2002) Oligosaccharide specificity of galectins: a search by frontal affinity chromatography. Biochim Biophys Acta 1572:232–254

Nakajima K, Kinoshita M, Matsushita N, Urashima T, Suzuki M, Suzuki A, Kakehi K. (2006) Capillary affinity electrophoresis using lectins for the analysis of milk oligosaccharide structure and its application to bovine colostrums oligosaccharides. Anal Biochem 348:105–114

Nakamura T, Amikawa S, Harada T, Saito T, Arai I, Urashima T (2001) Occurrence of an unusual phosphorylated N-acetyllactosamine in horse colostrum. Biochim Biophys Acta 1525:13–18

Pelletier I, Hashidate T, Urashima T, Nishi, N, Nakamura T, Futai M, Arata Y, Kasai K, Hirashima M, Hirabayashi J, Sato S (2003) Specific recognition of *Leishmania major* Poly-beta-galactosyl epitopes by Galactin-9. J Biol Chem 278:22223–22230

Uemura Y, Asakuma S, Nakamura T, Arai I, Taki M, Urashima T (2005) Occurrence of unique sialyl tetrasaccharide in colostrums of a bottlenose dolphin (*Tursiops truncates*) Biochim. Biophys Acta 1725:290–297

Urashima T, Nakamura T, Saito T (1997) Biological significance of milk oligosaccharides—Homology and heterogeneity of milk oligosaccharides among mammalian species. Milk Science 46:211–220

Urashima T, Yamashita T, Nakamura T, Arai I, Saito T, Derocher AE, Wiig, O (2000) Chemical characterization of milk oligosaccharides of the polar bear, *Ursus maritimus* Biochim Biophys Acta 1475:395–408

Urashima T, Saito T, Nakamura T, Messer M (2001) Oligosaccharides of milk and colostrums in non-human mammals. Glycoconj J 18:357–371

Section III
Sugar Chain Analysis by Mass Spectrometry

An XML Description of Carbohydrate Structures

Norihiro Kikuchi

Introduction

Several glycomics databases are being developed, and the numbers of glycan sequences stored in these are increasing because of improvements in high-throughput carbohydrate sequencing technologies such as mass spectrometry. However, the formats used to record glycan sequences differ between databases. The differences in the data formats hamper access and exchange of data, and prevent the bioinformatics community from developing novel software for glycomics. The complexity of carbohydrate sequences also makes it difficult to define a common language to represent them, and the development of bioinformatics tools for glycomics has not progressed.

Concept

Extensible Markup Language (XML) is playing an increasingly important role in the exchange of a wide variety of data on the Web, and has been applied to the distribution of biological data, for example, MIAME and XEMBL. Recently, we developed the carbohydrate sequence markup language CabosML; this was the first proposed XML data format for carbohydrate structure and enables standardization and distribution of carbohydrate sequences (Kikuchi et al. 2005). CabosML defines five features:

1. Description of the types of monosaccharides.
2. Description of the types of linkages and anomers.
3. Description of oligosaccharides containing branched structures.
4. Description of modifications.
5. Description of repeating and cyclic structures.

CabosML defines a carbohydrate sequence as a tree structure in which the monosaccharide at the reducing end is represented as the root node of a "g:Carb_structure" element in the XML document. The connections between adjacent monosaccharides are represented as parent–child relationships in the document: a parent element represents a monosaccharide on the reducing end-side, and a child element represents a monosaccharide on the non-reducing end-side. The principle of this approach can be illustrated using the carbohydrate structure of the N-Glycan core, Man5GlcNAc2:

Bioscience Group, Mitsui Knowledge Industry Co., Ltd., 7-14, Hitotsubashi SI bldg., 3-26, Kandanishiki-cho, Chiyoda-ku, Tokyo 101-0054, Japan
Phone: +81-3-5259-6570, Fax: +81-3-5280-2737
E-mail: kikuchi-norihiro@mki.co.jp

```
<g:Glyco xmlns:g="http://bio.mki.co.jp/glycoinformatics/2003">
  <g:Carb_ID/>
  <g:Carb_structure>
    <g:MS name="GlcNAc" >
      <g:MS link="1-4" anom="b" name="GlcNAc" >
        <g:MS link="1-4" anom="b" name="Man" >
          <g:MS link="1-3" anom="a" name="Man" />
          <g:MS link="1-6" anom="a" name="Man" />
        </g:MS>
      </g:MS>
    </g:MS>
  </g:Carb_structure>
</g:Glyco>
```

where the element "g:MS" represents the monosaccharide. The types of monosaccharides, anomers, and linkages are represented using the attributes "name", "anom," and "link", respectively. Modifications are defined as an element, "g:MOD", that has attributes of position of the modification on the monosaccharide. The following example shows the CabosML description of 6-sulfo sialyl Lewis X.

```
<g:Glyco xmlns:g="http://bio.mki.co.jp/glycoinformatics/2003">
  <g:Carb_ID/>
  <g:Carb_structure>
    <g:MS name="GlcNAc" >
      <g:MS link="1-3" anom="a" name="Fuc" />
      <g:MS link="1-4" anom="b" name="Gal" >
        <g:MS link="2-3" anom="a" name="Neu5Ac" />
        <g:MOD pos="6" name="S"/>
      </g:MS>
    </g:MS>
  </g:Carb_structure>
</g:Glyco>
```

In addition to modifications, CabosML can represent carbohydrates containing cyclic or repeating structures. The elements "g:COS" and "g:ROS" represent cyclic and repeating structures, respectively. The number of repeats is expressed using the attribute "num".

Application: Development of Database and Software Using CabosML

Carbohydrate sequences were described using CabosML and stored in the XML database. We developed a Web interface to insert, edit, delete, and search these sequences (Fig. 1). The depicted structures can be automatically converted into XML formats, and the interface enables users to search for particular structures. We also developed a software that calculates the theoretical m/z values of all possible ions fragmented from oligosaccharide sequences described by CabosML. The database and software were integrated into CabosDB (Carbohydrate sequencing database, see Chapter by N. Kikuchi in *Experimental Glycoscience: Glycobiology*).

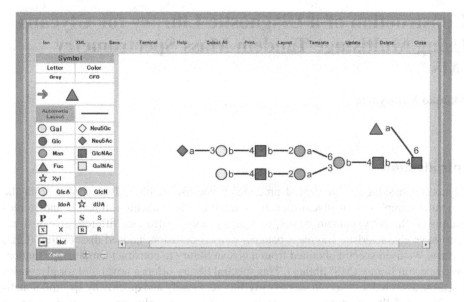

Fig. 1 A glycan editor. Users can draw a structure to register and search carbohydrate sequences

References

Kikuchi N, Kameyama A, Nakaya S, Ito H, Sato T, Shikanai T, Takahashi Y, Narimatsu H (2005) The carbohydrate sequence markup language (CabosML): an XML description of carbohydrate structures. Bioinformatics 21(8):1717–1718

Construction of a Diagnostic Library for Glycans Using Multistage Tandem Mass Spectrometry (MS^n)

Akihiko Kameyama

Introduction

Glycomics has lagged far behind proteomics because of difficulties arising from the structural complexity of glycans, such as variations in branching, linkage, and stereochemistry. Recently, tandem mass spectrometry has revealed that different glycan structures have characteristic fragment patterns in their collision-induced dissociation (CID) spectra. We used spectra obtained from a glycan library to construct a multistage tandem mass spectral library (MS^n library). This spectral library offers a novel tool for glycomics research, as it enables users to identify glycans very easily and quickly by spectral matching. The MS^n library stores triplicate spectra of standard glycans, and information on samples and experimental conditions in a relational database. The strategy of rapid identification is based on comparison of the signal intensity profiles of the spectra obtained by MS^n of the analyte with a library of mass spectral data (Kameyama et al. 2005).

Procedure

Materials

1. Sample solution: approximately 1 μM glycan solution in double distilled water.
2. Calibrant solution: peptide calibration standard II (Bruker Daltonics, Billerica, MA, USA) in 200 μL of 0.1% TFA in 50% acetonitrile.
3. Matrix solution: 10 mg/mL of 2,5-dihydroxybenzoic acid (2,5-DHB, proteomics grade, Wako, Japan) in 30% ethanol.
4. Recrystallization solution: 99.5% ethanol (HPLC grade, Wako, Japan).
5. Collision gas: argon

Methods

1. These instructions assume the use of a MALDI-QIT-TOF mass spectrometer (MS) equipped with an Ultra Cooling Kit (AXIMA-QIT; Shimadzu, Kyoto, Japan) and Kompact Version 2.5.1 operating software (Kratos Analytical Ltd., Manchester, UK).

Research Center for Medical Glycoscience, National Institute of Advanced Industrial Science and Technology (AIST), 1-1-1 Umezono, Tsukuba, Ibaraki 305-8568, Japan
Phone: +81-29-861-3200, Fax: +81-29-861-3201
E-mail: aki-kameyama@aist.go.jp

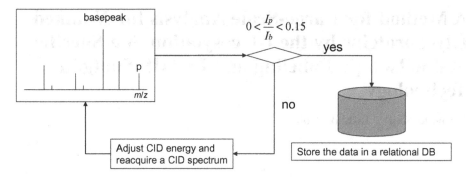

Fig. 1 Logic flow of evaluation of a valid CID spectrum for an MS^n spectral library. I_b: intensity of base peak, I_p: intensity of parent ion

2. Place 0.5 µL of the sample solution on a target plate (specular surface stainless steel plate, 2 mm diameter) and allow to dry. Next, cover the dried analyte on the target plate with 0.5 µL of the matrix solution and allow to dry. Finally, recrystallize the dried material by adding 0.15 µL of 99.5% ethanol to the matrix–analyte mixture on the target plate.

3. Acquire MS spectra of the glycans in the positive mode.

4. Acquire MS^2 spectra of sodium adduct ions derived from the glycans. The MS^2 spectra of the same ion should be acquired three times from different wells on the target plate. For the acquisition of CID spectra, adjust the CID energy so that the intensity of parent ions almost disappears. If the intensities of parent ions in the spectra are more than 15% of the base peak, the spectra must be discarded and reacquired with a larger CID energy (Fig. 1). Use an automatic acquisition function with the regular raster that governs the laser shot patterns (see *Comment 1*).

5. Acquire MS^3 spectra of all the major fragment ions in the MS^2 spectra in the same manner as described above for MS^2 acquisition.

6. Output the data including *m/z* values and their intensities in these CID spectra. Match with the corresponding glycan ID (see *Comment 2*), and input these data to a relational database, RDBMS Oracle9i (Oracle Corporation Japan, Tokyo, Japan).

Comments

1. CID energy adjustment and automatic acquisition are essential for reproducibility of fragment pattern.
2. Glycan IDs are linked to the corresponding data in the glycan structure database (DB) constructed independently (Kikuchi et al. 2005).

References

Kameyama A, Kikuchi N, Nakaya S, Ito H, Sato T, Shikanai T, Takahashi Y, Takahashi K, Narimatsu H (2005) A strategy for identification of oligosaccharide structures using observational multistage mass spectral library. Anal Chem 77: 4719–4725

Kikuchi N., Kameyama A., Nakaya S., Ito H., Sato T., Shikanai T., Takahashi Y., Narimatsu H (2005) The carbohydrate sequence markup language (CabosML): an XML description of carbohydrate structures. Bioinformatics 21:1717–1718

A Method for Large-Scale Analysis for *N*-linked Glycoproteins by the Glycosylation Site-Specific Stable Isotope-Labeling and LC/MS Shotgun Technology

Hiroyuki Kaji[1,2]**, Toshiaki Isobe**[1]

Introduction

Genome information constitutes a catalog of genes and proteins of an organism and allows genome-wide expression profiling of the transcriptome and proteome. A large body of evidence, however, suggests that many biological events are regulated not only by gene expression but also by post-translational modifications of proteins, such as glycosylation. Thus, the analysis of protein glycosylation is one of the most important issues of current proteomics research. We introduce here our strategy for large-scale "glycoproteome" analysis based on liquid chromatography-coupled mass spectrometry (LC/MS) shotgun technology.

Principle

The isotope-coded glycosylation site-specific tagging (IGOT) of glycopeptides, coupled with LC/MS-based identification of the tagged peptides, is designed for the large-scale analysis of *N*-linked glycoproteins in the complex biological samples. The steps of this approach are (1) lectin column-mediated affinity capture of glycopeptides generated by protease digestion of protein mixtures; (2) purification of the enriched glycopeptides by hydrophilic interaction chromatography; (3) peptide-N-glycanase (PNGase)-mediated incorporation of a stable isotope tag, ^{18}O, specifically at the *N*-glycosylation site; and (4) identification of ^{18}O-tagged peptides by LC/MS-based proteomics technology. The outline of this approach is illustrated in Fig. 1. The original methodology for IGOT (Kaji et al. 2003) and its detailed protocol (Kaji et al. 2006) have been published elsewhere.

[1]Department of Chemistry, Graduate School of Science and Engineering, Tokyo Metropolitan University, Minami-osawa 1-1, Hachioji, Tokyo 192-0397, Japan

[2]Research Center for Medical Glycoscience, National Institute of Advanced Industrial Science and Technology (AIST), Central 2-12, Umezono 1-1-1, Tsukuba, Ibaraki 305-8568, Japan
Phone: +81-29-861-5070, Fax: +81-29-861-2744
E-mail: kaji-rcmg@aist.go.jp

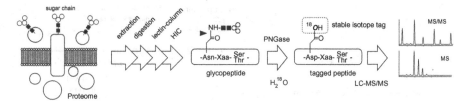

Fig. 1 Strategy for large-scale identification of *N*-linked glycoproteins by IGOT-LC-MS/MS. The steps include lectin column-mediated affinity capture and HIC purification of glycopeptides from the protease digests of protein mixture, PNGase-mediated incorporation of the stable isotope tag, ^{18}O, at the *N*-glycosylation site, and identification of the ^{18}O-tagged peptides by LC-MS/MS

Preparation of Glycopeptides from the Protease Digests of Protein Mixture

Our method of the glycoproteome analysis utilizes LC/MS shotgun technology, an advanced proteomics technology widely applied for large-scale protein analysis in complex biological mixtures. In this method, a protein mixture is digested directly with a protease such as trypsin, without protein separation, and the resulting extremely complex peptide mixture is analyzed by LC/MS. We considered that this technology would be suitable for glycoproteome analysis, because it would be applicable to the identification of many glycoproteins integrated into the membranes and might also clarify the problem of glycoproteins' heterogeneity due to multiple forms of oligosaccharide attached to the polypeptide chain. To collect glycopeptides specifically and comprehensively from highly complex peptide mixtures, we used multiple lectin columns with distinct binding specificity for non-reducing end saccharides, such as Con A (Man), RCA120 (Gal/Lac), WGA (GlcNAc), SSA (Sia), and AAL (Fuc), and purified the collected glycopeptides by hydrophilic interaction chromatography (HIC) on Sepharose CL-4B (Wada et al. 2004). We found that the HIC significantly improved the efficiency of glycopeptide identification.

Peptide-N-glycanase (PNGase)-mediated Stable Isotope Tagging of *N*-glycopeptide

The analysis of glycopeptides by tandem mass spectrometry following collision-induced dissociation (CID) preferentially generated a series of fragment ions derived from dissociation of glycosyl bonds rather than peptide bonds. This inhibited the assignment of the peptide by the shotgun approach. Thus, we introduced an experimental step to remove sugar chains attached to the polypeptide chain before the shotgun analysis. The *N*-linked glycans are removed from the polypeptide chain by digestion with PNGase regardless of the glycan structure. This reaction converts Asn at the sugar attachment site to Asp, thereby increasing the polypeptide mass by 1 Da. In principle, this could serve as a mass tag to identify a formerly glycosylated peptide; however, this chemical conversion must be distinguished from the potential deamidation reaction of Asn, which occurs frequently in vivo and in vitro (deamidation of Asn also increases the peptide mass by 1 Da). Our IGOT strategy was developed to solve this problem; only glycopeptides are selectively labeled with a stable isotope, ^{18}O, by performing the PNGase digestion of glycopeptides in $H_2{}^{18}$O (Fig. 1). This strategy also distinguishes glycopeptides from non-glycosylated

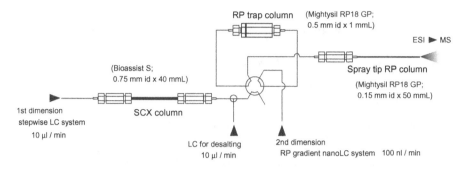

Fig. 2 Flow diagram of 2-dimensional nanoflow LC system. The system is composed of two-independent LCs equipped with a strong cation-exchange (*SCX*) column and a reversed-phase (*RP*) column, which are interfaced to perform automated LC/MS analysis

peptides in the lectin affinity-captured glycopeptide fraction, which may be contaminated due to non-specific binding to the affinity column.

Identification of *N*-glycopeptides by LC/MS Shotgun Analysis

The shotgun analysis of the resulting isotope-labeled peptides allowed large-scale identification of the glycoproteins with simultaneous determination of the glycosylated site. Our 2-dimensional nanoflow LC system is illustrated in Fig. 2. The system is composed of two-independent LCs equipped with an ion-exchange column and a reversed-phase (RP) column, which are combined with an electric column-switching valve and an RP-based solvent desalting unit to perform fully automated LC/MS analysis. The peptides are first separated on a strong cation exchange (SCX) column by stepwise elution with 10 buffers containing different KCl concentrations, and each fraction is further separated on an RP column by linear gradient elution. The eluted peptides are immediately applied to MS via an electrospray ionization (ESI) interface. To enhance the sensitivity of the analysis, the LC system is highly miniaturized (Natsume et al. 2002); it is equipped with a frit-less fused silica capillary spray-tip RP column (id = 150 μm, L = 50 mm, Fig. 2) and a sprit-less, direct nano-flow system composed of a high-precision nano-flow pump and a unique diffusion-mediated gradient device (ReNCon). The system can separate thousands of peptides in a few micrograms of sample within 24 h.

Each MS/MS spectrum is assigned to a particular peptide in the genome/protein sequence database by sequential data processing using the software "Masslynx" (Waters-Micromass) and "Mascot" (Matrix Science) according to the manufacturer's recommendations with several additional parameters such as error tolerance, fixed modification of Cys (e.g., carbamoylmethylation), and ^{18}O-containing Asp (for glycopeptide identification by IGOT, +3 Da). The large quantity of search results on glycopeptides are extracted and aligned by the software STEM (Shinkawa et al. 2005), which outputs the result file in tab-separated format that can be browsed using Excel (Microsoft).

Application and Future Perspective

By the IGOT-LC-MS/MS method described here using three types of lectin columns; Con A, wheat germ agglutinin (WGA), and worm galactose-binding lectin 6, we identified ~1,400 *N*-glycosylated sites on ~800 proteins from crude extracts of the nematode

C. elegans (Kaji et al. 2007). Because the method is compatible with quantitative proteomics technologies such as metabolic and chemical stable isotope labeling, it is applicable to the analysis of glycoproteome dynamics accompanying various biological events. Since protein glycosylation is strongly linked to physiology as evidenced by growing number of human diseases with defects in glycoconjugate assembly and processing, it will also be useful to search for the protein biomarkers for diagnosis and drug discovery.

Procedure for Sample Preparation of IGOT-LC/MS Analysis

1. Dissolve the protein source materials (whole organism, tissue, cells, organelles, or other biological materials) in extraction buffer (0.5 M Tris–HCl, pH 8.0, containing 6 M guanidine-HCl) by use of an appropriate equipment.
2. Add an aliquot of 100 mg/ml DTT to the sample solution at DTT:protein = 1:1 (w/w), and reduce disulfide bonds for 2 h at room temperature under the nitrogen atmosphere.
3. For S-carbamoylmethylation of cysteine residues, add iodoacetoamide (2.5:1 w/w protein), mix, and leave the mixture for 2 h in the dark.
4. Dialyze the reaction mixture at 4°C against 100-volumes of 10 mM HEPES–NaOH buffer, pH 7.5, to remove salts and excess reagents changing the buffer several times.
5. Add trypsin (1:50 w/w protein) into the solution and digest the proteins overnight at 37°C.
6. Load the peptide mixture onto the lectin column, wash the column, and elute glycopeptides with an elution buffer containing a mono/oligosaccharide appropriate for the lectin column.
7. Purify the peptide by HIC on Sepharose column, if necessary.
8. Evaporate the glycopeptides to dryness using a centrifugal vacuum concentrator.
9. Dissolve glycopeptides in PNGase buffer (10~20 µg/50~100 µl) prepared with $H_2^{18}O$. Add PNGase, dissolved in $H_2^{18}O$ at 1 mU/µl, at a final concentration of 1 mU/10 µg glycopeptide and incubate the mixture at 37°C overnight.
10. Analyze the ^{18}O-labeled peptide mixture using, e.g., an LC system connected to an electrospray-ionization tandem mass spectrometer.

References

Kaji H, Saito H, Yamauchi Y, Shinkawa T, Taoka M, Hirabayashi J, Kasai K, Takahashi N, Isobe T (2003) Lectin affinity capture, isotope-coded tagging and mass spectrometry to identify N-linked glycoproteins. Nat Biotechnol 21:667–672

Kaji H, Yamauchi Y, Takahashi N, Isobe T (2006) Mass spectrometric identification of N-linked glycopeptides using lectin-mediated affinity capture and glycosylation site-specific stable isotope tagging. Nat Protoc 1:3019–3027

Kaji H, Kamiie J, Kawakami H, Kido K, Yamauchi Y, Shinkawa T, Taoka M, Takahashi N, Isobe T (2007) Proteomics reveals N-linked glycoprotein diversity in *Caenorhabditis elegans* and suggests an atypical translocation mechanism for integral membrane proteins. Mol Cell Proteomics 6:2100–2109

Natsume T, Yamauchi Y, Nakayama H, Shinkawa T, Yanagida M, Takahashi N, Isobe T (2002) A direct nanoflow liquid chromatography-tandem mass spectrometry system for interaction proteomics. Anal Chem 74:4725–4734

Shinkawa T, Taoka M, Yamauchi Y, Ichimura T, Kaji H, Takahashi N, Isobe T (2005) STEM: a software tool for large-scale proteomic data analyses. J Proteome Res 4:1826–1831

Wada Y, Tajiri M, Yoshida S (2004) Hydrophilic affinity isolation and MALDI multiple-stage tandem mass spectrometry of glycopeptides for glycoproteomics. Anal Chem 76:6560–6565

Mass Spectrometry of Glycopeptides

Yoshinao Wada

Introduction

The key structural issues of glycoproteomics are protein identification, glycosylation site determination, and the glycan profiling at individual attachment sites (Tajiri et al. 2005). They are elucidated by tandem mass spectrometry of the glycopeptides isolated from tryptic digests of glycoproteins in a rapid and sensitive manner. Electrospray ionization (ESI) or matrix-assisted laser desorption/ionization (MALDI) mass spectrometry is employed (Fig. 1). In case of MALDI, profiling of oligosaccharides is achieved by the linear time-of-flight (TOF) mode more appropriately than the reflectron mode. Robust and reliable identification of proteins and the determination of the attachment sites often require multiple-stage tandem mass spectrometry (Wada et al. 2004).

Enrichment of Glycopeptides by Sepharose

1. Digest reductively alkylated proteins with trypsin in 0.05 M Tris–HCl, pH 8.0.
2. Mix 100 µg digest with a 15 µL packed volume of Sepharose CL4B or Superose-12 (GE Healthcare) in 1 mL of organic solvent (1-butanol/ethanol/H_2O, 5:1:1, v/v), and incubate for 45 min at room temperature with gentle shake.
3. Wash the gel with organic solvent three times.
4. Incubate the gel in 0.5 mL of aqueous solvent (ethanol/H_2O, 1:1, v/v), for 30 min, and recover the solution phase.
5. Evaporate the solvent by a Speed-Vac.
6. Reversed phase HPLC, if separation of individual glycopeptides or purity check is required.

Comments

Incubation with an excessive amount of gel increases contamination of peptides into the glycopeptide fraction.

Mass Spectrometry

1. Dissolve glycopeptides (0.1–1 µM) and 2,5-dihydroxybenzoic acid (10 mg/mL) in acetonitrile/H_2O/trifluoroacetic acid (50/50/0.1, v/v/v) separately.
2. Mix each 1 µL on MALDI sample target and dry.
3. Insert the sample target into MALDI mass spectrometer.
4. MS/MS database search.

Osaka Medical Center and Research Institute for Maternal and Child Health,
840 Murodo-cho, Izumi, Osaka 594-1101, Japan
Phone: +81-725-57-4105, Fax: +81-725-57-3021
E-mail: waday@mch.pref.osaka.jp

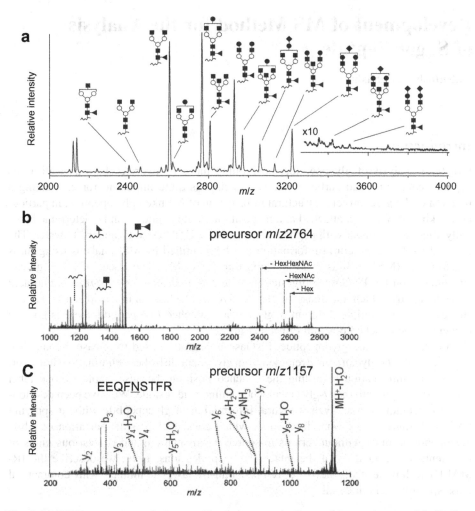

Fig. 1 MALDI mass spectra of glycopeptides from a tryptic digest of human immunoglobulin-G. **a** Mass spectrum taken in linear TOF mode. The glycan profile with the relative abundance of each oligosaccharide is obtained. **b** MS/MS spectrum for the ions at m/z 2764. The product ions indicating the molecular mass of peptide backbone sequence are observed in the low mass region. **c** MS/MS/MS for the ions at m/z 1157. Product ion search with this mass spectrum indicates amino acid sequence EEQF-NSTFR from immunoglobulin-G subclass-2 (IgG2). *filled triangle* fucose; *filled square* N-acetylglucosamine; *open circle* mannose; *filled circle* galactose; *open square* N-acetylneuraminic acid

Comment

ESI in the LC-MS setting is also effective for this purpose. Alternatively, glycopeptides are dissolved at 1 µM in acetonitrile/H_2O/formic acid (70/30/0.1, v/v/v) and directly infused into the ESI ion source for analysis.

References

Tajiri M, Yoshida S, Wada Y (2005) Differential analysis of site-specific glycans on plasma and cellular fibronectins: application of a hydrophilic affinity method for glycopeptide enrichment. Glycobiology 15:1332–1340

Wada Y, Tajiri M, Yoshida S (2004) Hydrophilic affinity isolation and MALDI multiple-stage tandem mass spectrometry of glycopeptides for glycoproteomics. Anal Chem 76:6560–6565

Development of MS Method for the Analysis of Sugar Peptide

Katsutoshi Takahashi

Introduction

Structural analysis of glycopeptide, positioning the carbohydrate chain as well as its peptide sequencing and carbohydrate sequencing, is quite important for elucidating a function of the glycoprotein. Structural information of N-linked glycopeptide, in particular the kind of N-glycan attached and its position on the peptide, can be determined partially after the treatment with peptide-N-glycosidase F (PNGase F) under ^{18}O water. The released N-glycan structural information can be identified by MS^n analysis using mass spectrometer (MS) such as quadrupol ion trap (QIT) MS. The traces of the N-glycan attachments on the PNGase F digested peptide can be detected by MS/MS amino acid sequencing, by which the mass of the N-glycan attached amino acid shifts +2 Da. However, when multiple different N-glycans are attached to a peptide, there is no way to identify the attached position of each N-glycan.

In case of O-linked glycopeptide, PNGase F cannot be used to release the attached O-glycans. The O-glycan can be released from glycopeptide by beta-elimination; however, it loses the information regarding the attached position, because stable isotope label cannot be brought into the O-glycan attached amino acid residue. We have been developing MS techniques to analyze structural information of glycopeptide without applying PNGase F digestion or beta-elimination, which consist of (i) soft ionization of labile glycopeptides intact without serious in-source fragmentation, and (ii) various kinds of fragmentation techniques of the intact glycopeptides ions. Here, we describe the IR-MALDI technique using newly developed mid-IR lasers combined with commercial mass spectrometers in detail.

Instrumentation

Two different special mid-infrared wavelength tunable nanosecond pulse lasers were newly developed in collaboration with Cyber Laser Inc. (Tokyo, Japan) and combined with commercial mass spectrometers such as MALDI-Q-FTMS and MALDI-QIT-TOFMS to achieve MS^n analysis of labile molecular ions generated by IR-MALDI soft-ionization.

Both mid-IR light sources consisted of a pump laser and an optical parametric oscillators (OPO) cavity. The pump laser was a diode pumped Nd:YAG laser, which was

Research Institute of Instrumentation Frontier (RIIF), National Institute of Advanced Industrial Science and Technology (AIST), Aomi 2-42, Koto-ku, Tokyo 135-0064, Japan
Phone: +81-3-3599-8039, Fax: +81-3-3599-8082
E-mail: takahashi-k@aist.go.jp

compact, efficient, and completely air cooled system. The laser was pulsed by using Electro-Optical Q-switch. One of the two mid-IR lasers had the OPO cavity in which there were multi-grating periodically poled lithium niobate (PPLN) crystals, and it emitted laser pulse stronger than 150 µJ/pulse at 100 Hz repetition rate at the whole wavelength tunable range (2.6–4.0 µm). This laser was combined with MALDI-Q-FTMS (APEX III, Bruker Daltonics, Billerica, MA, USA). The laser pulse was focused onto the MALDI sample plate with AR-coated CaF_2 lens of f = 200 mm to make 200 µm diameter spot.

In the other mid-IR laser, the OPO cavity had a newly developed 2-mm thick MgO: Periodically Poled Stoichiometric Lithium Tantalate (PPSLT) crystal. We obtained laser pulse over 0.67 mJ at the whole wavelength tunable range (2.8–3.2 µm), less than 20 ns pulse width, with easy operation by using full PC control. The laser pulse was focused onto the MALDI sample plate inside the commercial MALDI-QIT-TOFMS (AXIMA-QIT, Shimadzu Corp., Kyoto, Japan) at 400 µm diameter spot, to achieve laser-fluence of 0.58 J/cm^2 under 800 µJ/pulse laser output; its power was enough to ionize PA-labeled sugar molecules by IR-MALDI using thiourea or DHBA as matrix.

Materials and Analysis

Molecular structures of the compounds analyzed in this paper were listed in Table 1. The synthesized O-linked glycopeptide was kindly supplied from RIKEN (Saitama, Japan) and MALDI-FTMS analysis was carried out. The glycopeptides were dissolved in MilliQ water. The sample was mixed with two different kinds of matrix solutions; one was 80 mg/mL thiourea (Merck Ltd.) solution in MilliQ water and another was 10 mg/mL DHBA (Merck Ltd.) solution in 50%v/v MeOH/MilliQ water, in sample/matrix sol. v/v = 1:4. The sample/DHBA mixture was dropped onto the MALDI sample plate and dried at room temperature, whereas the sample/thiourea mixture was dried under vacuum chamber at room temperature. Figure 1 shows the mass spectra of UV-MALDI-FTMS analysis of sample/DHBA co-crystal and IR-MALDI-FTMS analysis of sample/thiourea co-crystal. With the 3 µm mid-IR laser pulse, which has little longer wavelength than that of Er.YAG laser, drastically decreased in-source fragmentation of the labile glycopeptide (Takada et al. 2004). In-source fragmentation of MALDI produced ion is thought to be caused by the excess energy applied during ionization process and to increase inside the MS^n capable ion-trap type mass spectrometer; such fragmentation

Table 1 Molecular structure of the compounds

Compound	Structure
Glycopeptide (RIKEN) H4.4	Galβ3- GalNAc | PSDTPILPQ
	Manα3-Manα6 | GlcNAcβ4-Manβ4-GlcNAcβ4-Glc NAc-PA | Galβ4-GlcNAcβ4-Manα3 | GlcNAcβ2

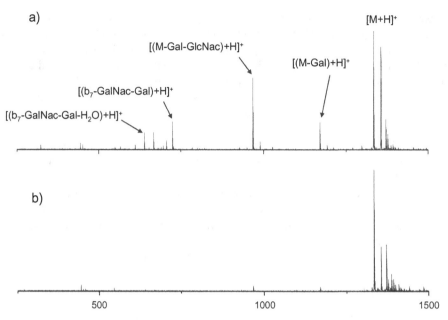

Fig. 1 UV/IR-MALDI-FTMS mass spectra of synthesized O-linked glycopeptide. **a** UV (N$_2$-laser) MALDI-FTMS mass spectrum of the glycopeptide was shown. Sample aqueous solution was mixed with 10 mg/mL DHBA in 50% MeOH/H$_2$O (1:4 v/v) and then dropped onto MALDI target followed by drying at room temperature. Four 80 μJ/pulse N$_2$-laser shots were used to acquire single spectrum. **b** IR (3.03 μm) MALDI-FTMS mass spectrum of the glycopeptides was shown. Sample aqueous solution was mixed with 80 mg/mL thiourea aqueous solution (1:4 v/v), dropped and dried in vacuum chamber. 300 shots of 200 μJ IR-laser (λ = 3.03 mm) pulses were used to acquire single spectrum

makes the sensitivity of the analysis low, and sometimes makes the MS^n of molecular ion impossible.

IR-MALDI-QIT-TOFMS analysis of labile glycopeptides also showed a very weak in-source fragmentation (data not shown), in which thiourea worked fined as IR-MALDI matrix. In Fig. 2, mass spectra of labile complex type PA-glycan molecule were shown. Both IR-MALDI-QIT-TOFMS and IR-MALDI-FTMS analyses showed less in-source fragmentation of the PA-glycan, while in both analyses only low-intensity molecular peak was found. In both cases, the in-source fragmentation level was almost same or less than that found in UV-MALDI-TOFMS (see Fig. 2c) even though the analysis time was 1,000-fold longer in MALDI-QIT-TOFMS and MALDI-FTMS. It should be noted that the sample consumption during the IR-MALDI analysis of this PA-glycan was almost the same as the UV-MALDI analysis. The IR-MALDI produced protonated or metal adduct of molecular ion can be fragmented further inside mass spectrometer to give structural information (Fukuyama et al. 2004; Budnik et al. 2005). MS^n analysis using IR-MALDI-QIT-TOFMS was also efficient in structural analysis of underivatized sulfated oligosaccharides to give their partial sequence information (Takahashi and Minamisawa 2007).

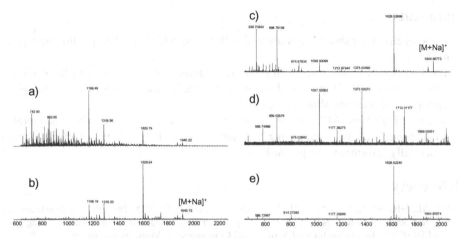

Fig. 2 UV/IR-MALDI mass spectra of labile complex type PA-glycan were shown. A 10 mg/mL DHBA dissolved in 50% MeOH/H₂O was used as UV-MALDI matrix solution, and an 80 mg/mL thiourea aqueous solution was used as IR-MALDI matrix solution. The 1:4 sample/DHBA mixture droplet was dried at room temperature for UV(N₂-laser) MALDI, whereas 1:4 sample/thiourea mixture droplet was vacuum dried at room temperature for IR (3.03 μm) MALDI experiments. **a** UV-MALDI-QIT-TOFMS spectrum. **b** IR-MALDI-QIT-TOFMS spectrum. **c** UV-MALDI-TOFMS spectrum. **d** UV-MALDI-FTMS spectrum. **e** IR-MALDI-FTMS spectrum

MS^n of Glycans or Glycoconjugates Using IR-MALDI-QIT-TOFMS

MS^n analysis of glycans or glycoconjugates are performed using IR-MALDI-QIT-TOFMS.

1. Prior to the analysis, confirm the wavelength and pulse energy of the IR-LASER to be 3.0 μm and 650–800 μJ/pulse, respectively.
2. Make 80 mg/mL thiourea aqueous solution as the matrix solution.
3. Dilute sample with pure water to appropriate concentration.
4. Mix sample solution and the matrix solution in v/v = 1:4 and vortex it for several seconds.
5. Drop 1 μL aliquots of the sample/matrix mixture onto the MALDI sample plate.
6. Put the MALDI sample plate in the vacuum chamber and dry the dropped aliquots at room temperature under vacuum.
7. Put the MALDI sample plate in the IR-MALDI-QIT-TOFMS ionization chamber.
8. Initiate the IR-LASER pulse and determine appropriate LASER power for IR-MALDI ionization, which depends on the sample.
9. Accumulate MS spectrum.
10. Isolate desired parent ion and perform MS^2.
11. Determine appropriate collision energy to obtain good MS^2 spectrum.
12. Accumulate MS^2 spectrum.
13. Isolate desired product ion to perform MS^3 analysis.
14. Repeat above operations 11–13 to obtain MS^n spectra.

Comments

1. Mass spectrometer should be calibrated with Mw standards, such as peptide-mix, prior to the analysis.
2. The IR-LASER power affects on both the ablation amount of the sample/matrix co-crystal and the MALDI produced ion intensity. The lowest LASER power which gives desirable ion intensity should be chosen.
3. In case of glycans, MS^2 of the parent molecular ion frequently gives dehydrated product ion. In many cases, further MS^3 analysis of such dehydrated product ion gives structurally informative fragment ions.

References

Budnik AB, Jebanathirajah AJ, Steen H, Costello EC, O'Connor BP (2005) Tandem excitation technique for sequencing biomolecules with labile post-translational modifications in Vibrational Cooling (VC) MALDI FTMS. In: Proceedings of 53rd ASMS Conference on Mass Spectrometry

Fukuyama Y, Wada Y, Yamazaki Y, Ojima N, Yamada M, Tanaka K (2004) A new analytical method for glycoprotein structure analysis using MALDI-QIT-TOFMS: an application to ribonuclease B. J Mass Spectrom Soc Jpn 52(6):328

Takada Y, Sumiyoshi T, Sekita H, Fukui K, Takahashi K (2004) Wavelength tunable mid-IR nanosecond light generation from optical parametric oscillator for IR-MALDI. In: Proceedings of 52nd ASMS Conference on Mass Spectrometry

Takahashi K, Minamisawa T (2007) Structural analysis of underivatized sulfated oligosaccharides by using IR-MALDI-QIT-TOFMS. In: Proceedings of 55th ASMS Conference on Mass Spectrometry

Calculations for Saccharides by the Use of MS Data

Kazuhiko Fukui

Introduction

High-sensitivity and high-throughput analysis by mass spectrometry (MS) is a powerful method for analyzing the structural complexity of oligo- and polysaccharides. Elucidating the fragmentation mechanisms in oligosaccharides using theoretical calculations is useful in analyzing experimentally obtained MS spectra. Empirical and/or theoretical rules for the fragmentation of oligosaccharides should be of considerable assistance in the development of novel tools for the structural analysis of oligosaccharides for use in glycomics. However, no rules that are applicable to large-scale glycomics exist for the fragmentation of oligosaccharides. This prompted us to explore, by theoretical calculation, the reasons why sodium-adduct ions of oligosaccharides produce certain characteristic fragment patterns. The fragmentation mechanisms could be useful in developing fragmentation rules, which would then assist in the development of a method for analyzing MS spectra.

Results

We have developed a Web-based tool named Structural Glycomics CALculations (SGCAL). SGCAL is capable of building a 3D structure from oligosaccharide sequence information and of visualizing the calculated results and the experimental MS data, thereby supporting investigations on correlations between the structure of oligosaccharides and their fragmentation patterns (Fig. 1). All collision-induced dissociation (CID) spectra were obtained from sodium adduct ions by using an MALDI-QIT-TOF mass spectrometer (AXIMA-QIT; Shimadzu). The observed fragment ions were recorded as lists of peaks in SGCAL. Theoretical calculations were performed for the oligosaccharides, and the results were compared with those obtained experimentally to provide information on structure–reactivity relationships. Several computational methods, such as molecular mechanics (MM3) calculations, semi-empirical (PM3/PM5) calculations, and electron orbital (HF and B3LYP) calculations, were used to interpret and analyze the experimentally observed fragmentation ions. Details of the principles of the experimental methods and calculations are given elsewhere (Fukui et al. 2006). SGCAL also has a mass-analysis function for searching through calculated results and experimental data: when mass values with an allowable mass range are entered into a search box, the mass-analysis function will retrieve lists of parent and fragment ions that meet the search criteria.

Computational Biology Research Center (CBRC), National Institute of Advanced Industrial Science and Technology (AIST), 2-41-6 Aomi, Koto-ku, Tokyo 135-0064, Japan
Phone: +81-3-3599-8060, Fax: +81-3-3599-8081
E-mail: k-fukui@aist.go.jp

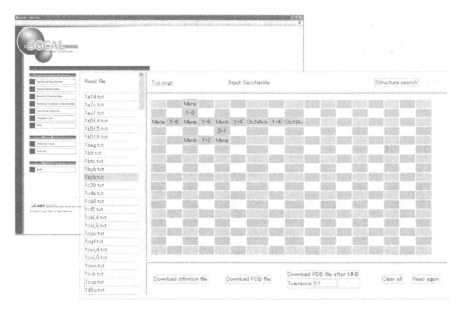

Fig. 1 SGCAL mainly consists of the following three functions: Fragmentation analysis, Mass analysis and Model creation. SGCAL is accessible at http://sgcal.cbrc.jp/

Theoretical calculations using computational chemistry-based methods can be useful in the analysis of experimentally obtained mass spectral data and for assigning oligosaccharide fragment ions. Theoretical rules for the fragmentation of oligosaccharides can be a novel tool for the structural analysis of oligosaccharides in glycomics (Yamagaki et al. 2006).

References

Fukui K, Kameyama A, Mukai Y, Takahashi K, Ikeda N, Akiyama Y, Narimatsu H (2006) A computational study of the structure–reactivity relationships in Na-adduct oligosaccharides in the collision-induced dissociation reactions. Carbohydr Res 341:624–633

Yamagaki T, Fukui K, Tachibana K (2006) Analysis of glycosyl bond cleavage and related isotope effects in collision-induced dissociation quadrupole/time-of-flight mass spectrometry of isomeric trehaloses. Anal Chem 78:1015–1022

Section IV
Analysis of the Three-dimensional Structure of Sugar Chains

Molecular Modeling of Oligosugar Structures

Masaji Ishiguro

Introduction

Structural data of oligosugars are essential for understanding their functions in molecular recognition in sugar–protein interactions. Further design of sugar mimics and non-sugar derivatives based on three-dimensional structures of oligosugars is one of the most important subjects in drug design.

Conformational analysis of oligosugars is studied in two ways. One is their solution structures, and the other is their biopolymer-bound conformations. Three-dimensional solution structures of oligosugars are constructed by use of distance information (NOE data) obtained by NMR measurements. The energy-optimized structures of oligosugars can be estimated by conformational analysis on glicosidic bonds. A number of combinations of the glycosidic bonds in oligosugars result in various kinds of their structures (Cioci et al. 2004; Imberty and Perez 2000).

For modeling of the structure-optimized oligosugars, each glycosidic bond is systematically analyzed by molecular mechanics and molecular dynamics calculations, and low-energy conformations are stored as database for dimeric sugar structures. Thus, the dimeric sugar structures in the database are retrieved and combined to construct initial oligosugar structures. Then, initial structures are energy-optimized by molecular dynamics/minimization calculations. The energy-optimized structures are classified into a cluster of low-energy conformations. Conformations of small oligosugars can be calculated in aqueous conditions with explicit water molecules, although it becomes hard to calculate conformations of larger oligosugars in aqueous solution conditions.

Complex structures of proteins and oligosugar ligands are constructed by docking the ligands into the ligand-binding cleft of proteins with distance information (NOE) between the ligands and the proteins obtained from NMR measurements (NOE data) or by use of docking programs which evaluate the complement of their shapes and electrostatics. Due to the flexibility of glycosidic bonds, each monomeric sugar moiety can be docked into the binding cleft, and the conformations of the glycosidic bonds are systematically searched at the cleft. Then, all glycosidic bonds are connected, and the initial conformations are energy-minimized and then structure-optimized in aqueous conditions by use of molecular mechanics and molecular dynamics calculations. In the case of glycotransferase, particular aromatic amino acid residues at the substrate-binding site are involved in the recognition of sugar moiety and determine the selectivity of the mode of the glycosidic bond (Kakuda et al. 2004; Shiba et al. 2006). For further analysis of sugar-transfer mechanism of the enzyme, quantum mechanics/molecular dynamics (QM/MD) method can be applied to understanding the reaction process.

Division of Computational and Biofunctional Research, Suntory Institute for Bioorganic Research, Shimamoto, Osaka 618-8503, Japan
Phone: +81-75-962-6028, Fax: +81-75-962-2115
E-mail: ishiguro@sunbor.or.jp

Concept

Recent studies on conformations of oligosugars have suggested that oligosugars have rather regular stable structures, and they take some stable conformations when they form complex structures with host biopolymers. It is shown that the interactions of sugars at protein surface are rather weak, and thus a wider protein surface and larger chain of sugars are required for a stable interaction. This implies that the simulation of oligosugar–protein interactions is expected to afford new aspects of molecular recognition.

References

Cioci G, Revet A, Koca J et al (2004) Conformational analysis of complex oligosaccharides: the CICADA approach to the uromodulin O-glycans. Carbohydr Res 339:949–959

Imberty A, Perez S (2000) Structure, conformation, and dynamics of bioactive oligosaccharides: the theoretical approaches and experimental validations. Chem Rev 100:4567–4588

Kakuda S, Shiba T, Ishiguro M et al (2004) Structural basis for acceptor substrate recognition of a human glucuronyltransferase, GlcAT-P, an enzyme critical in the biosynthesis of the carbohydrate epitope HNK-1. J Biol Chem 279:22693–22703

Shiba T, Kakuda S, Ishiguro M et al (2006) Crystal structure of GlcAT-S, a human glucuronyltransferase, involved in the biosynthesis of the HNK-1 carbohydrate epitope. Proteins 65:499–508

Prediction of Sugar-Binding Sites on Proteins

Tsuyoshi Shirai

Introduction

Computational techniques in prediction of protein–sugar (carbohydrate) interactions are required for the current biological studies. Although this field of bioinformatics is still immature as compared with that of protein–protein or protein–nucleic acid interactions, several methods have been developed as prompted by the increasing requirements.

The predictions of protein–sugar interactions are classified into two major categories. They are the predictions of post-translational glycosylation of glycoproteins and the non-covalent binding of sugars to proteins, mainly lectins and carbohydrate-processing enzymes. The former category is further classified into predictions of *N*- or *O*-glycosylation sites. The latter one consists of sugar-binding sites search on proteins and evaluation of stability of bound sugar molecules.

Glycosylation Site Prediction

Many of the proteins synthesized in higher organisms are glycosylated, and these post-translational modifications introduce considerable varieties in the properties of proteins such as solubility, sub-cellular localization, and interactions with other molecules. Therefore, detection of the glycosylation sites based on the sequences or three-dimensional structures of the nascent protein is important. These modifications are introduced via covalent attachment of sugars to Asn (*N*-glycosylation) or Ser/Thr (*O*-glycosylation) residues. Hydroxylysine can also serve for the acceptor of *O*-glycosylation.

The *N*-glycosylation sites are known to be readily detected with the sequence motif Asn-Xaa-(Ser/Thr) (Xaa stands for any amino acid). Most of the known sites have this consensus sequence, although not all of them are actually glycosylated.

O-glycosylation sites, on the other hand, rather show variable nature. There are at least seven types in *O*-glycosylation sites. Only ambiguous motifs, such as Xaa-Pro-Xaa-Xaa (one of the Xaa is glycosylated Ser or Thr), have been detected even for the most frequently observed and extensively studied mucin-type sites.

These problems have been mainly tackled by using the artificial neural network (ANN) or bio-based function neural network (BBFNN) techniques. The amino acid sequences of the experimentally confirmed glycosylation sites have been used for training the artificial neural networks, and predictions with nearly 90% accuracy have been achieved. The prediction methods of Gupta et al. for *N*-glycosylation site, Julenius et al. (2005), and Yang and Chou (2004) for mucin-type *O*-glycosylation site, and Johansen

Department of Bioscience, Nagahama Institute of Bio-Science and Technology, and
JST-BIRD, 1266 Tamura, Nagahama 526-0829, Japan
Phone: +81-749-64-8117, Fax: +81-749-64-8140
E-mail: t_shirai@nagahama-i-bio.ac.jp

Table 1 Applications for glycosylation site prediction

Application	Methods	Reference and availability
N-glycosylation site prediction (NetNGlyc)	ANN	Gupta et al. (unpublished). Available at http://www.cbs.dtu.dk/services/NetNGlyc
Mucin-type O-glycosylation site prediction (NetOGlyc)	ANN	Julenius et al. (2005). Available at http://www.cbs.dtu.dk/services/NetOGlyc
Mucin-type O-glycosylation site prediction	BBFNN	Yang and Chou (2004). Available at http://www.dcs.ex.uk/~zryang
Hydroxylysine-type O-glycosylation site prediction (NetGlycate)	ANN	Johansen et al. (2006). Available at http://www.cbs.dtu.dk/services/NetGlycate

et al. (2006) for hydroxylysine-type O-glycosylation site are currently available as the on-line applications (Table 1).

Protocol: N-Glycosylation Site Prediction with NetNGlyc Tool

1. Access NetNGlyc (http://www.cbs.dtu.dk/services/NetNGlyc) with a browser.
2. Paste or upload your amino acid sequence by either copy & pasting sequence to the box. "Paste a single sequence or several sequences in FASTA format into the field below," or entering the name of fasta format file into the box. "Submit a file in FASTA format directly from your local disk".
3. Press "Submit" button.
4. Prediction result will appear on screen as follows:

```
Name: Sequence  Length: 15
AFGHKNLTKILYTRE
.....N.........
(Threshold = 0.5)
-------------------------------------------------------------
SeqName  Position  Potential  Jury  N-Glyc
agreement  result
-------------------------------------------------------------
Sequence  6  NLTK  0.7747  (9/9)  +++
-------------------------------------------------------------
```

Sugar-binding Site Prediction

Sugar chains are the information-carrier molecules, in which their sequences and configurations encode their functions. The sugar chain information (glycocode) is generally read by proteins via a specific non-covalent binding of the sugar chains to the proteins. The protein–sugar interactions are used in various activities of life such as cell–cell recognition/adhesion, immune responses, and development. Predictions of the specificity, stability, and structures of protein–sugar complex are important in studying the biological roles of sugar molecules, and in applying sugar molecules for drug designs. Several computational applications for sugar-binding site predictions have been reported (Table 2).

Table 2 Applications for sugar-binding site prediction

Application	Methods	Reference and availability
Sugar-binding site prediction	Scoring with propensities, protrusion, and ASA of amino acids	Taroni et al. (2000)
Sugar-binding site, sugar-coordinates, and free energy prediction	AutoDock simulation with the modified potential energy function	Neumann et al. (2002) AutoDock is available for academic user at http://autodock.scripps.edu
Sugar-binding site and sugar-coordinates prediction	Scoring with empirical spatial distributions of protein atoms	Shionyu-Mitsuyama et al. (2003). Available on request to the author: t_shirai@nagahama-i-bio.ac.jp
Sugar-binding site, sugar-coordinates, and free energy prediction	AutoDock simulation	Laederach and Reilly (2003). AutoDock is available for academic user at http://autodock.scripps.edu
Sugar-binding site prediction	Scoring with propensities of amino acids	Malik, Ahmad (2007).

Taroni et al. (2000) have studied the preference of amino acids (propensity), shape of molecule (protrusion), and accessible surface area (ASA) of the sugar-binding sites on the three-dimensional structures of sugar–protein complexes. Then, they used the result of the observation for detecting the surface patches (cluster of amino acid residues) of sugar-binding sites. The overall accuracy of this prediction method was reported to be 65%. Shionyu-mitsuyama et al. (2003) employed the spatial distributions of protein atoms around the bound sugar moieties in the complex structures to derive an empirical score system for sugar-binding sites. This score system was applied to coordinate sugar moieties on the protein three-dimensional structures. Figure 1 shows an example of the prediction on *Erythrina corallodendron* lectin (PDB code 1lte). The predicted galactose with the best score (pGal1) was successfully allocated close to the co-crystallized galactose moiety. As the result of prediction tests on 50 protein structures, 59% of the sugar-binding sites were detected among top three predictions by using this method. Malik and Ahmad (2007) also used the propensity of amino acids to detect the residues used for sugar binding. Their score function can be applied for the primary structures of proteins. This method has provided the prediction performance of 87% sensitivity and 23% specificity.

Several docking simulations, which were specifically aimed at protein–sugar complex predictions, by using Autodock application, have been reported. Neumann et al. (2002) used the Autodock for a series of docking simulation of seven sugars to wheat germ agglutinin. They evaluated the binding-free energies of the sugars, and demonstrated that the estimated free energies had high correlation ($r = 0.96$) with the experimental IC_{50} values of the corresponding sugars. Laederach and Reilly (2003) have introduced a modified potential energy function, in which the terms of hydrogen-bond energy in solution and non-polar accessible surface area were added to the conventional function. They applied this function for the docking simulations of 17 inhibitors to the *Aspergillus niger* glucoamylase. The standard error of the predicted binding-free energy from the experimentally determined ones was shown to be 1.101 kcal/mol.

Fig. 1 Galactose-binding site prediction on *Erythrina corallodendron* lectin. pGal1~3 are the predicted galactose coordinates, and Galactose (co-crystallized) indicates the experimentally determined position of galactose

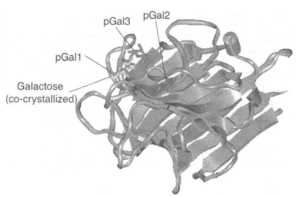

References

Julenius K et al (2005) Prediction, conservation analysis, and structural characterization of mammalian mucin-type O-glycosylation sites. Glycobiology 15:153–164

Laederach A, Reilly PJ (2003) Specific empirical free energy function for automated docking of carbohydrates to proteins. J Comput Chem 24:1748–1757

Malik A, Ahmad S (2007) Sequence and structural features of carbohydrate binding in proteins and assessment of predictability using a neural network. BMC Struct Biol 7:1

Johansen MB et al (2006) Analysis and prediction of mammalian protein glycation. Glycobiology 16:844–853

Neumann D et al (2002) Lectin-sugar interaction calculated versus experimental binding energies. Eur J Biochem 269:1518–1524

Shionyu-Mitsuyama C et al (2003) An empirical approach for structure-based prediction of carbohydrate-binding sites on proteins. Protein Eng 16:467–478

Taroni C et al (2000) Analysis and prediction of carbohydrate binding sites. Protein Eng 13:89–98

Yang ZR, Chou KC (2004) Predicting the linkage sites in glycoproteins using bio-basis function neural network. Bioinformatics 20:903–908

Section V
Analysis of Sugar–Protein Interactions

Frontal Affinity Chromatography: An Effective Analytical Tool for Protein–Sugar Interaction

Kenichi Kasai

Introduction

Frontal affinity chromatography (FAC) is currently the most powerful tool for studying molecular recognition performed by sugar-binding proteins, and actually the only method that enables detailed profiling of the specificity of sugar-binding proteins, because it has the following distinct merits that no other analytical procedure for molecular interaction has. First, FAC is suitable for the analysis of weak-binding phenomena characteristic to protein–sugar interactions from its principle, although almost all presently available procedures are applicable only to strongly interacting systems. Second, from practical viewpoints, consumption of rare oligosaccharides in FAC can be kept minimal because of its extremely high sensitivity (an amount of ~20 pmol oligosaccharide is sufficient for one measurement). Even the procedure having the highest sensitivity other than FAC probably requires ten times more sugar molecules. This is why FAC is actually the only available procedure for profiling sugar-binding proteins.

The principle and practice of FAC as a general analytical tool for biospecific interactions were established in the 1970s by Kasai et al. (Kasai and Ishii 1975). At its primitive stage, FAC was rather time-consuming and required a relatively large amount of analyte (Ohyama et al. 1985; Kasai et al. 1986). However, significant improvements in both instruments and materials have now made FAC a highly sensitive, accurate, rapid, stable, and convenient procedure for the determination of equilibrium constants between an immobilized ligand and a soluble analyte (Hirabayashi et al. 2000, 2003). In its application to the field of glycobiology, the use of a small column of immobilized lectin and fluorescent oligosaccharides (PA-sugars; structurally diverse oligosaccharides labeled with pyridylamino group are commercially available) has made FAC an extremely effective, sensitive, and high-throughput tool for profiling sugar-binding proteins in terms of their binding properties (both binding specificity and binding strength). It becomes easy to systematically determine K_d values of up to 100 kinds of PA-sugars for a given sugar-binding protein in a few days.

The remarkable characteristics of FAC are as follows: (1) the theoretical basis is simple and straightforward because this method treats only an equilibrium state (more precisely, a dynamic equilibrium state) established between an immobilized ligand and a soluble analyte; (2) FAC fits to relatively weak interactions in contrast to almost all currently available techniques; (3) neither special equipment nor sophisticated skill is needed for operation because the simplest elution mode, isocratic elution, is applied

Biochemistry, Faculty of Pharmaceutical Sciences, Teikyo University, Sagamiko, Kanagawa 229-0195, Japan
Phone: +81-42-685-3740, Fax: +81-42-685-3742
E-mail: kasai-k@pharm.teikyo-u.ac.jp

Fig. 1 Principle of frontal affinity chromatography

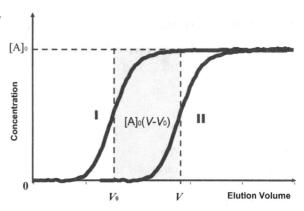

throughout the chromatographic runs; (4) an accurate elution volume can be determined easily by processing integrated multiple data points, thus minimizing the influence of noise in the measurement of signals and resulting in acquisition of reliable equilibrium constants (K_d values); (5) once a minimum set of basic parameters for a given column is established, K_d values for other analytes can be determined without knowing their exact concentrations. This is a remarkable advantage of FAC; (6) there is no need to use a high concentration of analytes for the purpose of facilitating complex formation and acquisition of larger signals even for weak interactions. The minimum concentration that allows drawing of an elution pattern is sufficient regardless of its affinity; (7) it is economically advantageous because only ordinary HPLC system is needed.

Principle

In FAC, a relatively large volume of dilute analyte solution is continuously applied to a small column packed with an affinity adsorbent (e.g., a PA-sugar solution is applied to an immobilized lectin column). It is important that a relatively weak affinity adsorbent that does not tightly bind the target analyte should be used. Unlike ordinary chromatography, the elution curve is composed of the front and the plateau as shown in Fig. 1. The amount of analyte molecules retarded during the passage through the column is equal to the area within the two elution curves: the right curve (II) being that of the analyte, and the left curve (I) being that of a reference substance having no affinity for the adsorbent. This area corresponds to the amount of the analyte molecules bound to the immobilized ligand in the column, and is equal to the rectangle, $[A]_0(V - V_0)$, where $[A]_0$ is the initial concentration of the analyte, V is the elution volume of the analyte, and V_0 is that of the reference substance. Therefore, $[A]_0(V - V_0)$ indicates the degree of saturation of the immobilized ligand, and is a function of the dissociation constant (K_d), the amount of immobilized ligand in the column (B_t), and $[A]_0$.

$$[A]_0(V - V_0) = \frac{B_t[A]_0}{[A]_0 + K_d}$$ Eq. 1

This equation is equivalent to the Michaelis–Menten equation of enzyme kinetics and in principle to Langmuir's adsorption isotherm. Equation 1 gives a hyperbolic curve, indicating that the column becomes saturated at an infinite concentration of analyte. B_t and K_d correspond to the maximum velocity and Michaelis constant, respectively, and can be determined by using one of the linear plots derived from Eq. 1.

If $[A]_0$ is set negligibly low in comparison with K_d , Eq. 1 can be simplified as follows:

$$(V - V_0) = B_t / K_d$$
$$\text{Eq.2}$$

This indicates that the extent of retardation is proportional to the reciprocal of K_d (i.e., association constant). Therefore, once the B_t value of a given affinity column is determined by concentration dependence analysis using an appropriate analyte, K_d values for other analytes can be determined by only one measurement of the V value provided that the analyte concentration is adequately low (e.g., lower than 1% of the K_d). This feature provides a great advantage from experimental viewpoints; that is, it is not needed to know the correct concentration of the analyte, as in the case of enzyme kinetics where enzyme concentration can be kept unknown. Moreover, even for a weakly interacting analyte (having a large K_d), concentration of the analyte solution can be kept as low as possible for detection, and results in the minimum consumption.

System and Procedure

Affinity adsorbent is usually prepared by immobilizing a sugar-binding protein to one of the commercially available activated carriers, e.g., NHS-activated Sepharose. It is noted that a relatively weak affinity adsorbent, which does not bind sugar molecules too tightly but allows their leakage, should be prepared, because we have to measure the elution volume of the front. As shown in Eq. 2, the extent of retardation of the elution front depends on B_t/K_d. Therefore, from practical viewpoints, it is recommended to control the concentration of immobilized protein between 1 and 10 times of presumed K_d values. For example, if a given sugar-binding protein has MW of 10,000 and presumable K_d values of around 10^{-5} M for PA-sugars, preferable immobilized protein concentrations are between 0.1 and 1 mg/ml.

An example of FAC system for the determination of K_d values of PA-sugars for sugar-binding proteins is shortly described below. It is easy to construct by using commercially available parts of HPLC. It is composed of a pump, an injector connected to a sample loop of relatively large volume (e.g., 2 ml), a small column (e.g., 4.0×10 mm^2, 126 µl) packed with an adsorbent to which a sugar-binding protein is immobilized, and a fluorescent detector. Because equilibrium states are sensitive to temperature change, the column is immersed in a water bath. Output from the fluorescence detector is transformed into digital signals and sent to a computer for data processing.

To the column, an appropriate buffer is continuously applied at a flow rate of 0.25 ml/min. The sample loop is filled with a PA-sugar solution at a concentration of ~10 nM, and then, the content is applied to the affinity column. Elution of PA-sugar is monitored with a fluorescence detector (excitation, 320 nm; emission, 400 nm). Because the column is automatically washed after the application of a PA-sugar solution is finished, it is necessary to wait only until the signals return to the base level before applying the next sample. Once the Bt value of a given column is determined by a concentration dependence experiment according to Eq. 1, K_d values of other PA-sugars can be determined one by one with only a single run for each by using an ~10 nM solution. Because it takes only about 10 min for a single run, determination of several tens of K_d values per day is not difficult, and profiling of a given sugar-binding protein in terms of its binding property is completed in a few days.

FAC has opened a door leading to deeper understanding of significance of sugar recognition in living systems. Detailed sugar-binding profiles of numbers of commercially available lectins have greatly augmented their utility as a research tool, and applications of FAC to a variety of newly found sugar-binding proteins are now greatly promoting elucidation of their true functions. Readers are encouraged to challenge FAC because it is not difficult at all and will provide you with extremely rich fruits.

References

Hirabayashi J, Arata Y, Kasai K (2000) Reinforcement of frontal affinity chromatography for effective analysis of lectin-oligosaccharide interactions. J Chromatogr 890:261–271

Hirabayashi J, Arata Y, Kasai K (2003) Frontal affinity chromatography as a tool for elucidation of sugar recognition properties of lectins. Methods Enzymol 362:353–368

Kasai K, Ishii S (1975) Quantitative analysis of affinity chromatography of trypsin. A new technique for investigation of proteinligand interaction. J Biochem 77:261–264

Kasai K, Nishikata M, Oda Y, Ishii S (1986) Frontal affinity chromatography. Theory for its application to studies on specific interaction of biomolecules. J Chromatogr 376:323–329

Ohyama Y, Kasai K, Nomoto H, Inoue Y (1985) Frontal affinity chromatography of ovalbumin glycoasparagines on a concanavalin A-Sepharose column. A quantitative study of the binding specificity of the lectin. J Biol Chem 269:6882–6887

Analyses of Sugar–Protein Interactions by NMR

Yoshiki Yamaguchi, Koichi Kato

Introduction

Analysis of sugar–protein interaction is an important step for elucidating the structure and function relationships of glycans. NMR spectroscopy serves an invaluable tool for this purpose because it is applicable for detecting weak sugar–protein interactions ($K_d \sim$ mM), identifying glycotope recognized by protein, characterizing conformation of oligosaccharide bound to protein, and determining the modes of interaction between oligosaccharide and protein in solution at atomic level (Fig. 1).

Concept

Upon sugar–protein interactions:

1. Mobilities of oligosaccharide molecules in solution (e.g., translational diffusion and rotational Brownian motion) are reduced.
2. Protons included in the glycotope and those located in the sugar-binding site of protein become spatially closer to one another.
3. Microenvironments surrounding the sugar-binding site are perturbed.
4. Conformation of oligosaccharide can be altered.

All these events can be detected by applying appropriate NMR techniques, which offer detailed insights into the sugar–protein interactions. For instance, transferred NOE (trNOE) experiment is useful for studying conformation of smaller oligosaccharide molecules weakly bound to larger proteins, on the basis of NOE-based distance information, even though the structural information of the protein is absent: An NOE between two protons in the bound ligand is transferred to the free molecule by chemical exchange process (Jiménes-Barbero and Peter 2006). In practice, the trNOE experiment works well for ligands bound to large proteins in fast exchange on the NMR timescale. TrNOE signals can be simply detected in 2D ^{1}H–^{1}H NOESY spectra.

Procedures for trNOE Experiment

1. Prepare a 0.1 mM protein solution and record the ^{1}H NMR spectrum in the absence of the ligand.
2. Record the ^{1}H NMR spectra of the protein solution typically with 0.2, 0.4, 0.6, 0.8, 1.0, 1.5 and 2.0 molar equivalents of oligosaccharide.
3. Check that the protein–ligand interaction is in fast exchange on the NMR timescale.

Graduate School of Pharmaceutical Sciences, Nagoya City University, Tanabe-dori 3-1, Mizuho-ku, Nagoya 467-8603, Japan
Phone: +81-52-836-3447, Fax: +81-52-836-3447
E-mail: kkato@phar.nagoya-cu.ac.jp

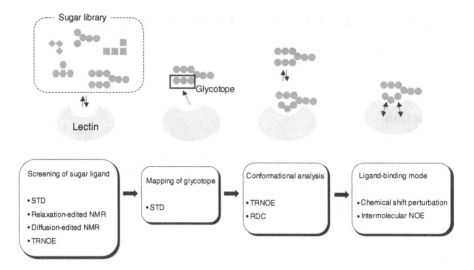

Fig. 1 A scheme of strategy for analyzing sugar–protein interactions

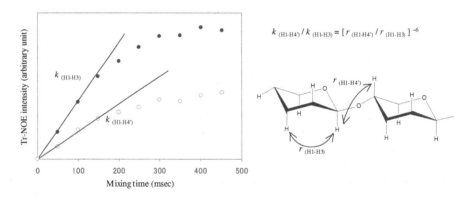

Fig. 2 Calculation of ^{1}H–^{1}H distances from the NOE build-up curve of intra- and inter-residue trNOE

4. Record the 2D NOESY spectra of protein–oligosaccharide complex (protein : oligo-saccharide = 1 : 5 to 1 : 20) at varying mixing times (e.g., 50, 100, 150, 200, 250, 300, 350 and 400 ms).
5. Plot the intensity of trNOE signals as a function of the mixing time (NOE build-up curve).
6. NOE build-up curve can be used to calculate interproton distances, provided intensity changes are proportional to cross-relaxation rates (Fig. 2).

Figure 3 shows a part of trNOESY spectrum for a sugar/lectin complex, which indicates that the conformation of the trimannose ligand is altered upon binding to the lectin.

For weakly binding sugar ligands, a conventional approach termed saturation transfer difference (STD)-NMR spectroscopy is available by which the glycotope of the ligand recognized by target protein can be identified (Mayer and Meyer 2001). This approach

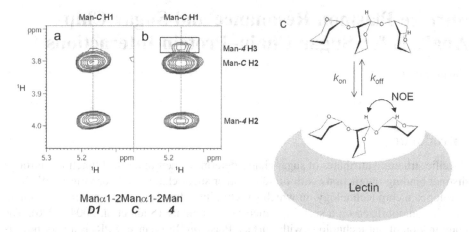

Fig. 3 A part of NOESY spectrum of Manα1-3Manα1-3Man (**a**) compared to the corresponding region of trNOESY spectrum of this trimmannose bound to the carbohydrate recognition domain of VIP36, an intracellular animal lectin (**b**). Interglycosidic NOE between Man-*C* H1 and Man-*4* H3 (*boxed*) is observed in the trNOESY spectrum but not in the NOESY spectrum, indicating that these protons are in a spatial proximity in the lectin-bound trimannose (**c**)

has also been applied for finding an oligosaccharide that is capable of interacting with a target protein from the oligosaccharide mixture.

Stable isotope labeling of sugar chain and/or its cognate protein greatly facilitates NMR analyses of sugar–protein interactions. Inspection of chemical shift perturbation observed for isotopically labeled protein upon interacting with oligosaccharide allows us to map sugar-binding site thereon (Mizushima et al. 2004). On the basis of inter-molecular NOE connectivities observed by stable-isotope-assisted NMR methods, the sugar-binding mode of proteins can be revealed at atomic resolution (Yamaguchi and Kato 2006). Residual dipolar coupling (RDCs) data, which is useful for determining the conformation of oligosaccharide in association with a protein, can be readily collected by use of ^{13}C-labeled oligosaccharides (Thompson et al. 2000).

The knowledge gained through the NMR analyses of sugar–protein interactions is crucial for designing new oligosaccharide-based therapeutic agents.

References

Jiménes-Barbero J, Peters T (2006) TR-NOE experiment to study carbohydrate–protein interactions. In: Jiménes-Barbero J, Peters T (eds) NMR spectroscopy of glycoconjugates. Wiley-VCH, Weinheim, pp 289–309

Mayer M, Meyer B (2001) Group epitope mapping by saturation transfer difference NMR to identify segments of a ligand in direct contact with a protein receptor. J Am Chem Soc 123:6108–6117

Mizushima T, Hirao T, Yoshida Y, Lee SJ, Chiba T, Iwai K, Yamaguchi Y, Kato K, Tsukihara T, Tanaka K (2004) Structural basis of sugar-recognizing ubiquitin ligase. Nature Struct Mol Biol 11:365–370

Thompson GS, Shimizu H, Homans SW, Donohue-Rolfe A (2000) Localization of the binding site for the oligosaccharide moiety of Gb3 on verotoxin 1 using NMR residual dipolar coupling measurements. Biochemistry 39:13153–13156

Yamaguchi Y, Kato K (2006) Structural glycobiology by stable-isotope-assisted NMR spectroscopy. In: Webb GM (ed) Modern magnetic resonance, vol 1. Springer, The Netherland, pp 219–225

Surface Plasmon Resonance and Sugar Chip Analysis for Sugar Chain–Protein Interactions

Yasuo Suda

Introduction

Specific structural attributes of sugar chains determine their biological functions, through distinct binding interactions with proteins, other sugar chains, cells or viruses. We have developed a chip technology, in which structurally defined sugar chains are two-dimensionally immobilized on a surface, which mimics nature (Suda et al. 2004, 2006). The combination of our technology with Surface Plasmon Resonance (SPR) is a very powerful tool for the real-time study of the specific interactions between biological molecules, and could be used as a high-throughput screening method or as a novel diagnosis technology, because the experiment can be done without any labeling of targets. Not only for regular SPR apparatus, but our technology can be applied also for the so-called SPR Imaging apparatus.

Protocols

1. Preparation of a Sugar Chip and SPR Experiment (An Example)

Glcα1-4Glc (maltose) was conjugated with the mono-valent linker molecule to prepare a ligand-conjugate (abbreviated Glcα1-4Glc-mono) for the immobilization of sugar chains on the gold-coated chip as reported (Suda et al. 2006), in which the glucose unit at the reducing end lost the sugar structure and worked as a hydrophilic spacer between the gold surface and the sugar chains. The synthesis of Glcα1-4Glc-mono was depicted in Fig. 1. An optimized reductive amination reaction and a well-designed linker compound (Koshida et al. 2001; Suda et al. 2006), in which aromatic amino group and the cyclic disulfide moiety were key units, were selected. SPR experiments were performed with SPR-670M (Moritex Corp., Yokohama, Japan) under the recommended manufacturer's guideline with a slight modification. Sensor chips used for SPR experiment were prepared as follows. Gold-coated chips (SUDx-Biotec Corp., Kobe, Japan) were soaked in 1 μM solution (methanol/water = 1/1, v/v) of the ligand-conjugate at room temperature for 2 h or overnight, followed by subsequent washing with methanol/water containing 0.05% tween-20, phosphate buffered saline (PBS) at pH 7.4 containing 0.05% tween-20, and PBS (pH 7.4). All the washing was done with a microwave for 5–30 min. The solvent for the binding experiment was PBS at pH 7.4, run at a flow rate of 15 μl/min at 25°C. Figure 2 shows an result of SPR experiment using Glcα1-4Glc-mono immobilized Sugar Chip for the binding of Concanavalin A. It was confirmed that no binding was observed

Department of Nanostructure and Advanced Materials, Graduate School of Science and Engineering, Kagoshima University, 1-21-40 Kohrimoto, Kagoshima, Japan
Phone: +81-99-285-8369, Fax: +81-99-285-8369
E-mail: ysuda@eng.kagoshima-u.ac.jp

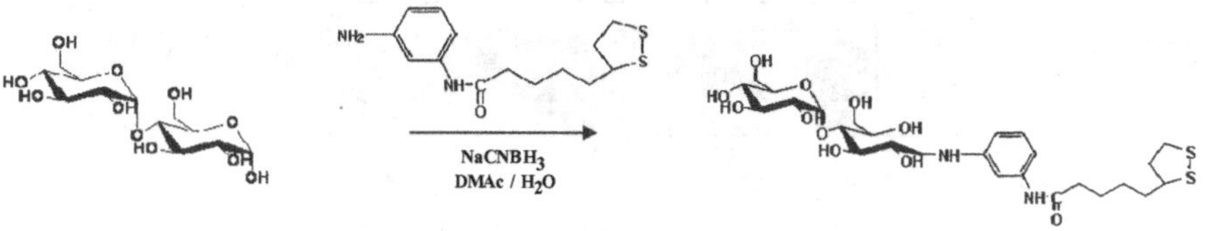

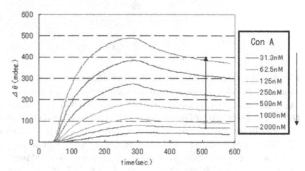

Fig. 1 Preparation of ligand-conjugate by optimized reductive amination reaction

Fig. 2 SPR sensor gram for the binding of Con A against Glcα1-4Glc-mono immobilized chip

when BSA was run on this chip. The sensor gram in Fig. 2 was just the plain data and not the subtracted one, which is regularly obtained by the subtraction of reference sensor gram from sample one. More than 50 kinds of Sugar Chips immobilized with a single sugar chain as prepared above are now commercially available from SUDx-Biotec.

2. Preparation of Array-Type Sugar Chip and SPR Imaging for the Binding of Lectins

For SPR imaging experiment, 1 μl of the aqueous solution containing 0.5% glycerol of the appropriate ligand-conjugates (500 μM) were spotted on the chip with automatic spotter (Toyobo, Osaka, Japan) according to the manual of the manufacturer. The spotted array-type chip was placed in a stand for 1 h at room temperature and washed sequentially with water, PBS containing 0.05% Tween-20 (PBS-T) and water, and dried at room temperature. The array-type Sugar Chip immobilized with various kinds of sugar chains (maximum 96 kinds) as described earlier was set on a prism with refraction oil in an SPR Imaging apparatus (MultiSPRinter™, Toyobo). The SPR measurements were performed according to the manual of the manufacturer using PBS-T as a running buffer at the flow rate of 100 μl/min at room temperature. Figure 3 shows the specificity of lectin binding for sugar chains immobilized on chip. The concentration of lectins was fixed at 2 μM in PBS-T. The contracted experiments using those techniques are now available in SUDx-Biotec.

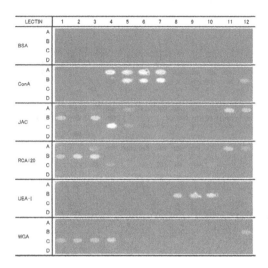

Fig. 3 SPR imaging of the binding of lectins against array-type Sugar Chip. Ligand-conjugates spotted at the address (designated in parenthesis) of chip are as follows: Glc-mono (*A1*); GlcNAc-mono (*A2*); Gal-mono (*A3*); Glcα1-4Glc-mono (*A4*); Glcα1-4Glcα1-4Glc-mono (*A5*); Glcα1-6Glc-mono (*A6*); Glcα1-6Glcα1-6Glc-mono (*A7*); Glcβ1-3Glcβ1-3Glc-mono (*A8*); Glcβ1-4Glc-mono (*A9*); Glcβ1-6Glc-mono (*A10*); Galα1-6Glc-mono (*A11*); Galα1-4Galβ1-4Glc-mono (*A12*); Galβ1-3GalNAcα1-6Glc-mono (*B1*); Galβ1-4GlcNAcβ1-6Glc-mono (*B2*); Galβ1-4Glc-mono (*B3*); Galβ1-4[Fucα1-2]GlcNAcβ1-3Galβ1-4Glc-mono (*B4*); Manα1-2Man-mono (*B5*); Manα1-3Manα1-4Man-mono (*B6*); Manα1-6Man-mono (*B7*); Fucα1-2Galβ1-4Glc-mono (*B8*); Fucα1-6Glc-mono (*B9*); Fucβ1-6Glc-mono (*B10*); Xylβ1-6Glc-mono (*B11*); GlcNAcα1-6Glc-mono (*B12*); GlcNAcβ1-4GlcNAc-mono (*C1*); GlcNAcβ1-6Glc-mono (*C2*); GlcNAcβ1-3Galβ1-4GlcNAcβ1-3Galβ1-4Glc-mono (*C3*); GalNAcα1-6Glc-mono (*C4*); GalNAcβ1-3Gal-mono (*C5*); NeuAcα2-3Galβ1-4GlcNAc-mono (*C7*); NeuAcα2-3Galβ1-3GlcNAcβ1-6Glc-mono (*C8*); NeuAcα2-3Galβ1-4GlcNAcβ1-6Glc-mono (*C9*); NeuAcα2-6Galβ1-4Glc-mono (*C10*); NeuAcα2-6Galβ1-3GlcNAcβ1-6Glc-mono (*C11*); NeuAcα2-3Galβ1-4GlcNAcβ1-6Glc-mono (*C12*); GlcNS6Sα1-4IdoA2Sβ1-6Glc-mono (*D1*); Heparin-mono (*D2*)

References

Suda Y et al (2004) WO 2004/022583 A1; WO 2004/022565 A1

Suda Y, Arano A, Fukui Y, Koshida S, Wakao M, Nishimura T, Kusumoto S, Sobel M (2006) Immobilization and clustering of structurally defined oligosaccharides for Sugar Chips: an improved method for surface plasmon resonance analysis of protein–carbohydrate interactions. Bioconjug Chem 17: 1125–1135

Koshida S, Suda Y, Arano A, Sobel M, Kusumoto S (2001) An efficient method for the assembly of sulfated oligosaccharides using reductive amination. Tetrahedron Lett 42:1293–1296

Analysis of Interactions Between Carbohydrates and Proteins Using Capillary Affinity Electrophoresis

Mitsuhiro Kinoshita,[1] **Kazuaki Kakehi**[2]

Introduction

Glycosylation is one of the most important post-translational events for proteins, and plays important roles for expression of various biological events such as cell–cell interaction. However, it should be noticed that glycans attached to the protein are not a primary gene product but the product from the biosynthesis by combination of various glycosyl-transferases, glycosidases, nucleotide donors, and nucleotide transporters. Thus, the glycans which attach to the core protein are highly heterogeneous in abundance, structure, and linkage position.

The analytical approach using capillary affinity electrophoresis (CAE) makes it possible to determine the post-translational modification of proteins with carbohydrates. The method is based on high-resolution separation of fluorescent-labeled carbohydrates by capillary electrophoresis with laser-induced fluorescent detection in the presence of carbohydrate-binding proteins (e.g., lectins). CAE is a powerful technique for simultaneous analysis of the interactions between glycans having similar structures and lectins even in complex mixtures with ultra high sensitivity.

Principle

We developed CAE for profiling complex mixtures of oligosaccharides derived from glycoconjugates (Nakajima et al. 2003, 2004, 2006; Kinoshita and Kakehi 2005). In CAE, the interactions are observed in free solution state, although most of the reported ones are based on the interactions between the carbohydrates and the immobilized lectins. In addition, it should be noted that the method is available for the analysis of the interaction between each carbohydrate in a mixture of glycans and a lectin. The principle of CAE is shown in Fig. 1.

At the initial step, a mixture of fluorescent-labeled carbohydrate chains (A, B and C) is analyzed by capillary electrophoresis in the absence of a lectin (Fig. 1a). In the following step, the same sample is analyzed in the presence of a lectin whose specificity is well established. When the lectin recognizes carbohydrate A (peak A in Fig. 1a), the peak is observed later due to the equilibrium based on the interaction with the lectin. On the contrary, carbohydrate C (peak C) does not show affinity to the lectin, and is observed

Faculty of Pharmaceutical Sciences, Kinki University, Kowakae 3-4-1, Higashi-Osaka, Osaka 577-8502, Japan
Phone: +81-6-6721-2332, Fax: +81-6-6721-2353
E-mail: [1]m-kino@phar.kindai.ac.jp, [2]k_kakehi@phar.kindai.ac.jp

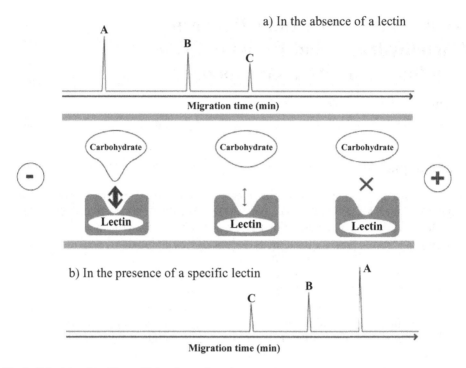

Fig. 1 Principle of capillary affinity electrophoresis

at the same migration time as that observed in the absence of the lectin. When carbohydrate B (peak B) shows weak binding to the lectin, the peak is observed slightly later. Thus, the electrophoretic behavior of the carbohydrate chains changes as shown in Fig. 1b. By repeating the procedures described above using an appropriate set of lectins, we can categorize all carbohydrate chains in the mixture.

As mentioned above, the most important point of CAE is that the binding specificity of each carbohydrate even in the complex mixture can be determined, and differences in binding specificities of glycans having similar structure are discriminated. After examining more than 30 lectins for CAE, we selected some lectins such as Con A, WGA and RCA that show obvious affinity to specific glycans and are stable in the analytical conditions.

Procedures

Preparation of 8-Aminopyrene-1,3,6-Trisulfonate (APTS)-Labeled Carbohydrates

1. To a sample of oligosaccharides (ca. 0.1–1 nmol), a solution (5 µl) of 100 mM APTS in 15% aqueous acetic acid is added.
2. A freshly prepared solution of 1 M NaBH₃CN in tetrahydrofuran (5 µl) is added to the mixture.
3. The mixture is overlaid with mineral oil (100 µl) to prevent evaporation of the reaction solvent to obtain good reproducibility.

4. The mixture is kept for 90 min at 55°C.
5. Water (200 μl) is added to the mixture. The fluorescent yellowish aqueous phase is collected, and applied on a column of Sephadex G-25 (1 cm i.d., 50 cm length) previously equilibrated with water.
6. The earlier eluted fluorescent fractions are collected and evaporated to dryness. The residue is dissolved in water (100 μl), and a portion (10 μl) is used for capillary affinity electrophoresis.

Capillary Affinity Electrophoresis of APTS-Labeled Glycans

1. CAE is performed with a capillary electrophoresis system equipped with a laser-induced fluorescence detector (Beckman P/ACE MDQ Glycoprotein system). An Ar-laser (ex. 488 nm) and a He–Cd laser (ex. 325 nm) are available for CAE analysis. Separation is performed using an inner surface modified capillary (typically a Beckman eCAP N-CHO capillary or a DB-1 capillary: 50 μm i.d., 10–30 cm effective length), because lectins are often adsorbed on a capillary surface during analysis. Lectins are dissolved in 100 mM Tris-acetate buffer (pH 7.4) at the concentration of 100 μM. Store the lectin solutions in aliquot at −20°C until use. All the lectin solutions are diluted with 100 mM Tris-acetate buffer (pH 7.4) to a specified concentration, and a portion (5 μl) of the solution is mixed with the running buffer (5 μl) containing 1% polyethylene glycol.
2. Prior to the analysis, the capillary is rinsed with 100 mM Tris-acetate buffer (pH 7.4) for 1 min at 20 p.s.i.
3. The capillary is filled with the same buffer containing 0.5% polyethylene glycol for 1 min at 20 p.s.i.
4. A sample solution containing fluorescent-labeled oligosaccharides is introduced for 10 s at 1.0 p.s.i.
5. Analysis is performed under constant voltage mode of 6 kV.
6. The capillary is washed with the same buffer containing 0.5% polyethylene glycol for 1 min at 20 p.s.i.
7. The capillary is rinsed with a lectin solution at the specified concentration in the same buffer containing 0.5% polyethylene glycol.
8. Analysis of the same glycan mixture is performed under constant voltage mode of 6 kV.
9. When necessary, the above procedures 6–8 are repeated at different concentrations of the lectin or for the different set of lectins.
10. The electropherograms observed in the presence of the lectin are compared with that observed in the absence of the lectin.

Applications

A typical example for CAE is shown in Fig. 2 using a mixture of asialo-N-glycans derived from human α1-acid glycoprotein (AGP) as model.

AGP contains five complex-type N-glycans as shown in the right panel of Fig. 2. I is a biantennary glycan, II and IV are tri- and tetra-antennary glycans, respectively. III and V are tri- and tetra-antennary glycans, respectively, having Lewis X residues by attaching a fucose residue. After labeling of these glycans with APTS, the mixture of the labeled glycans is analyzed by capillary electrophoresis. The left panel shows the migrations of

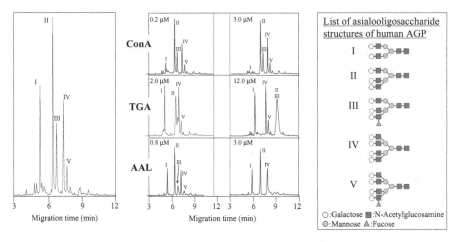

Fig. 2 Profiling of N-linked glycans derived from human α1-acid glycoprotein by capillary affinity electrophoresis. Analytical conditions: capillary, eCAP N-CHO capillary (30 cm total length, 10 cm effective length, 50 μm i.d.); running buffer, 100 mM Tris-acetate (pH 7.4) containing 0.5% PEG70000; applied voltage, 6 kV (reverse polarity); sample injection, pressure method (1.0 p.s.i., 10 s). Fluorescence detection was performed with a 520 nm emission filter by irradiating with an Ar laser-induced 488 nm light. *Left panel* is the profile of oligosaccharides from human AGP without lectin, and the *center panel* is those with lectins. Structures of oligosaccharides (**I–V**) are shown at the *right panel*

these five glycans in the absence of carbohydrate-binding protein (i.e., lectin). All glycans are well resolved within 10 min. It should be noticed that good resolutions are achieved between **II** and **III**, and **IV** and **V**. These sets of glycans are resolved by the difference of one fucose residue.

We can identify these glycans by CAE using the set of Con A (a lectin from Jack Bean), TGA (a lectin from tulip bulbs), and AAL (a lectin from *Aleuria aurantia*). Con A obviously decreases the peak intensity of biantennary glycan (**I**), and the peak of **I** disappears at 3.0 μM concentration of Con A. On the contrary, tri- and tetraantennary glycans (**II–V**) do not show any changes in the presence of Con A. Triantennary glycans (**II** and **III**) can be confirmed using the electrolyte containing TGA. At 2.0 μM of TGA, both triantennary glycans of **II** and **III** are observed slightly later. Finally, these peaks are fused into a broad peak and observed after those of tetraantennary glycans (**IV** and **V**) at 12.0 μM of TGA; **I**, **IV**, and **V** do not show any changes in the presence of TGA. Glycans having fucose residues attached to GlcNAc residue of lactosamine at one of the branches are determined using the electrolyte containing AAL. At 0.8 μM of AAL, peak intensities of **III** and **V** are obviously decreased, and these peaks are disappeared at 3.0 μM of AAL.

As described in the procedures above, we can easily identify the glycans from the set of binding data with lectins and migration times of the glycans. Time required for single CAE run is within 15 min. Therefore, we can characterize the glycome derived from biological samples such as cultured cells or serum in a few hours by CAE techniques.

References

Kinoshita M, Kakehi K (2005) Analysis of the interaction between hyaluronan and hyaluronan-binding proteins by capillary affinity electrophoresis: significance of hyaluronan molecular size on binding reaction. J Chromatogr B Analyt Technol Biomed Life Sci 816:289–295

Nakajima K, Oda Y, Kinoshita M, Kakehi K (2003) Capillary affinity electrophoresis for the screening of post-translational modification of proteins with carbohydrates. J Proteome Res 2:81–88

Nakajima K, Kinoshita M, Oda Y, Masuko T, Kaku H, Shibuya N, Kakehi K (2004) Screening method of carbohydrate-binding proteins in biological sources by capillary affinity electrophoresis and its application to determination of *Tulipa gesneriana* agglutinin in tulip bulbs. Glycobiology 14(9): 793–804

Nakajima K, Kinoshita M, Matsushita N, Urashima T, Suzuki M, Suzuki A, Kakehi K (2006) Capillary affinity electrophoresis using lectins for the analysis of milk oligosaccharide structure and its application to bovine colostrum oligosaccharides. Anal Biochem 348(1):105–114

Equilibrium Dialysis

Shunji Natsuka

Introduction

Since reported by Marrack and Smith (1932), the equilibrium dialysis method has been used as a technique for analyzing the association constant between molecules in liquid phase (Fig. 1). However, other methods have become more commonly used in recent years, as the equilibrium dialysis has two distinct disadvantages in terms of speed and relatively large volumes required.

In many of the techniques used for measurement of sugar chain–protein interactions these days, sugars or proteins are fixed on solid carriers. There is a theoretical weak point in such techniques which is the reason that comparison between the results from separate experiments is not simple, because the fixed density vastly influences the association. In particular, this property is marked in low-affinity systems such as sugar–lectin interactions. In many cases, the substances strongly associate at high-density fixation but do not show any interaction at low-density fixation. In contrast, equilibrium dialysis is an excellent method for measuring binding specificity, because it is theoretically simple, as both sugars and proteins are homogeneous in liquid phase.

Nowadays it is possible to measure the association constant with tens to hundreds of micrograms of proteins by equilibrium dialysis using improved microanalysis (Mega and Hase 1992). However, in order to compete with high-throughput devices, for example a microchip, the sensitivity of equilibrium dialysis has to be improved. In current methods, 100 µL of sample is required for analysis. If this volume can be reduced to 1 µL, this technique offers the possibility of becoming a method of measuring, superior to other techniques in terms of both sensitivity and convenience.

Procedures

1. Inject 50 µl of a lectin solution in 50 mM sodium phosphate buffer, pH 7.0 into a dialysis cell. (detail of the dialysis cell is described in Mega and Hase 1991)
2. Inject 50 µl of a ligand solution (ex.: PA-glycans) in 50 mM sodium phosphate buffer, pH 7.0 into another compartment of the dialysis cell.
3. Set at 4°C for 24 h without shaking.
4. Analyze a portion of the ligand solution with HPLC apparatus.

Department of Chemistry, Graduate School of Science, Osaka University,
1-1 Machikaneyama, Toyonaka, Osaka 560-0043, Japan
Phone: +81-6-6850-5381, Fax: +81-6-6850-5382
E-mail: natsuka@chem.sci.osaka-u.ac.jp

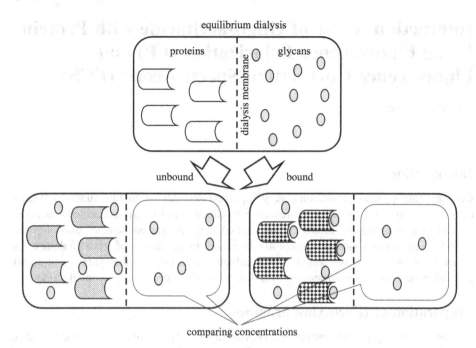

Fig. 1 Outline of equilibrium dialysis

References

Marrack JK, Smith FC (1932) Quantitative aspects of immunity reactions: The combination of antibodies with simple haptenes. Br J Exp Pathol 13:394–402

Mega T, Hase S (1991) Determination of lectin-sugar binding constants by microequilibrium dialysis coupled with high performance liquid chromatography. J Biochem 109:600–603

Interaction Assay of Oligosaccharide with Protein Using Fluorescence Polarization (FP) and Fluorescence Correlation Spectroscopy (FCS)

Toshiyuki Inazu

Introduction

Carbohydrate–protein interaction is very important in understanding the function of sugar chains. The immobilization or fluorescence modification of oligosaccharides was investigated to understand the biological interaction between various species and oligosaccharides. In many cases of immobilization and modification of carbohydrates, the structure of the glycan was found to be relatively small, because the synthesis of an oligosaccharide is not easy compared with that of an oligopeptide or oligonucleotide.

Preparation of Glycosylasparagines

Various kinds of glycosylasparagines (H-Asn(Sugar)-OH) are obtained by the peptidase digestion of glycopeptides from hen eggs (Tai et al. 1975; Seko et al. 1997). We obtained N-Fmoc (9-Fluorenylmethyloxycarbonyl) glycosylasparagine derivatives with high-mannose-type natural N-glycan from ovalbumin (Inazu and Mizuno 1999). The N-Fmoc glycosylasparagine derivatives having the disialo and asialo complex-type N-glycan were obtained from sialoglycopeptide (SGP) in egg yolk. Fmoc-Asn(Sugar)-OH was purified by HPLC using an octadecylsilica (ODS) column (Inazu and Mizuno 1999; Yamamoto et al. 2003). Several kinds of Fmoc-Asn(Sugar)-OH are presently available (Otsuka Chemical Co, Tokyo, Japan).

Interaction Assay Using Glycochip (Mizuno et al. 2004)

Fmoc-Asn(Sugar)-OH was immobilized in a 96-well plate, such as a Nunk CovaLink[TM] NH Module, on which the amino group was modified using 4-(4,6-dimethoxy-1,3,5-triazin-2-yl)-4-methylmorpholinium chloride (DMT-MM; Kunishima reagent), as shown in Fig. 1a. The sialo-complex-type oligosaccharide H-Asn(Sialo)-OH was immobilized using bis(sulfosuccinimidyl)suberate (BS^3). The binding assay of the horseradish peroxydase (HRP)-labeled lectins, ConA, RCA120, DSA, and WGA, to the glycochip and the subsequent inhibition of a binding event of M6-ConA were performed by ELISA as shown in Fig. 2a, b. These results showed the limitation of the glycochip (solid

Department of Applied Chemistry, School of Engineering, and Institute of Glycotechnology,
Tokai University, 1117 Kitakaname, Hiratsuka, Kanagawa 259-1292, Japan
Phone: +81-463-58-1211, Fax: +81-463-50-2012
E-mail: inz@keyaki.cc.u-tokai.ac.jp

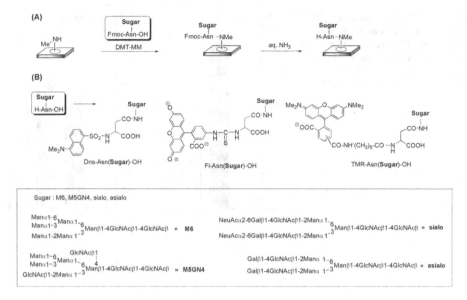

Fig. 1 A Immobilization of oligosaccharide in a 96-well plate; **B** introduction of fluorescence groups to glycosylasparagine derivatives

phase glyco-tool) caused by a nonspecific interaction between a plastic plate and a protein. Although strong interactions, such as M6 and ConA, were easily detected by these kinds of glycochips, detection of weak interactions, such as sialo and WGA, is difficult.

Interaction Assay Using Fluorescence Polarization and Fluorescence Correlation Spectroscopy (Mizuno et al. 2004)

The carbohydrate–protein interactions in the solution assay system were carried out using glycosylasparagine derivatives, FP and FCS. Fluorescence groups, dansyl (Dns), fluorescein (Fl), and tetramethylrhodamine (TMR), were introduced into the amino group of glycosylasparagines (Fig. 1b). The interaction of Dns-Asn(M6)-OH with ConA was analyzed using FP. This result showed significant binding of ConA with Dns-Asn(M6)-OH. We calculated the K_d (dissociation constant) for a ConA of 0.34 mM. The FCS assay using TMR-Asn(M6)-OH also gave a K_d for a ConA of 0.56 mM. The inhibition assay was also demonstrated by FP and FCS.

TMR-Asn(Sugar)-OH (M6, sialo, asialo, M5GN4) and lectins (ConA, RCA120, WGA) were mixed. After 10-min incubation at room temperature, the diffusion time was measured by FCS, providing the interaction ratio shown in Fig. 2c. For example, the interaction of ConA with M6 was stronger (76%) than its interaction with sialo (41%), and no interaction was observed with M6-RCA120. From the perspective of the assay of the weaker interaction between an oligosaccharide and a protein, the assay in a solution system, such as FP and FCS, appears to be superior to the interaction assay by a solid-phase glyco-tool such as glycochip.

Fig. 2 A Binding assay of HRP-ConA with glycochips; **B** binding assay of HRP-RCA 120 with glycochips; **C** interaction of TMR-glycosylasparagines (5 nM) with lectins (5 μM) measured by FCS

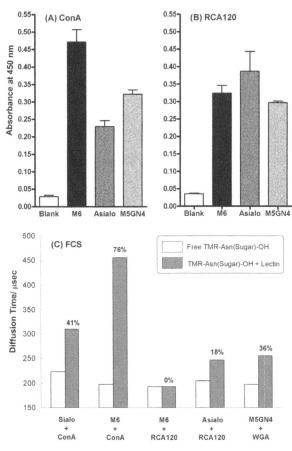

Protocol

Introduction of Fluorescence Groups to the Glycosyl Asparagine (H-Asn(Sugar)-OH)

To the solution of H-Asn(Sugar)-OH and 3 equiv. of *N, N*-diisopropylethylamine in DMF was added 1.5–2.0 equiv. of fluorescence reagent, such as Dns-Cl, FITC, and TMR-OSu, and stirred at room temperature for 18 h. HPLC (ODS column) purification gave the desired compound in 64–97% yield.

Measurement of Interaction of Lectins and Sugar

TMR-Asn(M6)-OH and Con A were mixed. After 10-min incubation at room temperature, its diffusion time was measured by use of MF20TM (Olympus Corporation). The final concentration of TMR-Asn(M6)-OH was 5 nM, and Con A was 5 μM. The result was shown in Fig. 2.

References

Inazu T, Mizuno M (1999) Preparation of sugar chain-linked asparagine active ester derivative and the synthetic intermediate. Jpn Kokai Tokkyo Koho: JP11-255807

Mizuno M, Noguchi M, Imai M, Motoyoshi T, Inazu T (2004) Interaction assay of oligosaccharide with lectins using glysocylasparagine. Bioorg Med Chem Lett 14:485

Seko A, Koketsu M, Nishizono M, Enoki Y, Ibrahim HR, Juneja LR, Kim M, Tamamoto T (1997) Occurrence of a sialylglycopeptide and free sialylglycans in hen's egg yolk. Biochim Biophys Acta 1335:23

Tai T, Yamashita K, Ogata-Arakawa M, Koide N, Muramatsu T, Iwashita S, Inoue Y, Kobata A (1975) Structural studies of two ovalbumin glycopeptides in relation to the endo-β-N-acetylglucosaminidase specificity. J Biol Chem 250:8569

Yamamoto N, Ohmori Y, Sakakibara T, Sasaki K, Juneja LR, Kajihara Y (2003) Solid-phase synthesis of sialylglycopeptides through selective esterification of the sialic acid residues of an Asn-linked complex-type sialyloligosaccharide. Angew Chem Int Ed 42:2537

Development of Neoglycoconjugate Probes and Detection of Lectins

Haruko Ogawa, Keiko Nakagawa

Introduction

Glycoconjugates are involved in essential biological processes, and interference with glycosylation is pathological and sometimes lethal in many organisms. Because the biological functions of glycans are exhibited primarily in their interactions with carbohydrate-binding proteins (lectins or receptors), which recognize individual glycoconjugates, we have developed various types of probes to detect lectins and analyze glycan interactions with high sensitivity. Commercially available glycoprobes are prepared by organic synthesis, a method that is suitable for large-scale production but requires large amounts of glycans and specialized synthesis steps. However, preparation of probes by glycobiologists in the laboratory needs to be technically feasible and applicable to only minute amounts of glycans or glycoconjugates, which can be purified from biological sources to provide a probe of adequate sensitivity for detection and to prepare affinity adsorbents. Our probes are prepared by derivatizing natural glycans or glycoconjugates by nondenaturing aqueous-phase reactions. In this report, we describe the development of neoglycoconjugate systems for sensitive lectin screening and interaction analyses. Examples of probes for lectin screening from animal, plant and fungal sources are shown (Ueda et al. 1999; Matsumoto et al. 2001; Ogawa et al. 2003; Satoh et al. 2004).

Principles of Neoglycoconjugate Probe Preparation

Carbohydrate-binding proteins have conventionally been studied using probes that have glycan chains or glycoproteins labeled with a functional tag such as biotin, a fluorescent chromogen or an epitope for immunoenzymatic detection (Fig. 1a). Glycans are usually conjugated with a hydrophobic group of a protein or lipid to immobilize them on the solid phase, and these are called hybridglycoproteins or neoglycolipids (Chai et al. 2003) (Fig. 1b, c). For affinity adsorbents, glycosaminoglycans (GAGs) are directly immobilized onto gel beads (Fig. 1d).

Neoproteoglycans and Hybrid Glycoproteins

Neoproteoglycans (neoPGs) are developed by conjugating acidic or neutral glycans with proteins to immobilize GAGs or oligosaccharides effectively on solid substrates. NeoPGs

Graduate School of Humanities and Sciences and the Glycoscience Institute, Ochanomizu University, 2-1-1 Otsuka, Bunkyo-ku, Tokyo 112-8610, Japan
Phone: +81-3-5978-5343, Fax: +81-3-5978-5343
E-mail: ogawa.haruko@ocha.ac.jp

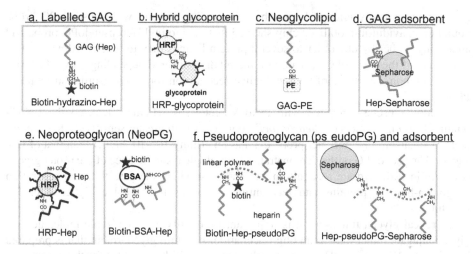

Fig. 1 Conventinally prepared glycoconjugate probes and adsorbents and those developed for detection of lectins and interaction analyses

detect binding with high sensitivity; for example, heparin is coupled with bovine serum albumin (BSA) and labeled with biotin to obtain biotin-BSA-heparin to detect heparin-binding proteins (Fig. 1e, right; Chai et al. 2003). A series of neoPG probes containing colominic acid, fucoidin, heparin and pectic acid were used to screen lectins from extracts of 16 cultivable mushrooms in combination with various hybrid horseradish peroxidase (HRP)-conjugated glycoproteins having different glycan types (Fig. 1b). The specificities of the lectins in the mushroom extracts were classified according to the reaction pattern of the probe on screening by dot-blot assay, and one lectin was purified (Matsumoto et al. 2001). Using the neoPG probe prepared by coupling heparin with HRP (Fig. 1e, left), the interactions with *Psathyrella velutina* lectin or vitronectin are analyzed by solid-phase assays: ELISA, affinity chromatography, and Western blotting (Ueda et al. 1999; Ogawa et al. 2003).

Pseudoproteoglycans (PseudoPGs)

A proteoglycan (PG) is a supermolecule consisting of one or more GAG chains attached to a core protein that binds link proteins and hyaluronans. PGs have signaling roles and cell modulatory functions in the extracellular matrix and at the cell surface. Some conventional probes, such as biotin-hydrazino-Hep and neoPGs, react with biological ligands other than those of PGs because these probes are dissimilar from the higher-order structure of natural PGs. Pseudoproteoglycans (pseudoPGs) simulate the structure of a PG monomer with more than one GAG side chain attached to a linear polymer strand, such as poly-L-lysine or linear polyacrylamide, and are biotinylated using NHS-biotin. Then, the remaining amino groups are blocked by *N*-acetylation to obtain biotinyl GAG-pseudoPG probes for use as probes and affinity adsorbents to search for and locate PG-binding substances in biological materials (Fig. 1f; Satoh et al. 2004). Binding studies indicate that heparin-binding proteins react differently to biotinyl pseudoPG and biotin hydrazine-Hep on Western blotting and affinity chromatography. Remarkably, several proteins in rat brain extracts bound specifically to either, but not both, of the probes, indicating that the Hep-binding proteins recognize a higher-order structure of PG. Surface plasmon resonance studies of known heparin-binding proteins with a pseudoPG probe bound to an avidin-immobilized chip showed that the biotinyl Hep-pseudoPG probe had an affinity higher by one order to several protein ligands than to biotin-hydrazino-Hep. The effect of the pseudoPG structure on the binding differed depending on the heparin-binding protein, and some of them exhibited recognition for a higher-order structure of PG.

PseudoPG probes prepared with lacto-N-neotetraose (LNnT) detected the LNnT-binding lectin in extracts of *Pleurocybella porrigens* with greater sensitivity than a commercial sugar-biotinyl polymer probe that has an affinity 10^2–10^5 higher than that of free sugars. The pseudoPG probe demonstrates the biological significance of the proteoglycan structure in carbohydrate recognition and can be used for the detection and separation of glycan-binding proteins and recognition of events that involve higher-order PG structures.

To reveal glycan functions, it is important to identify a recognizable molecule as a glycoconjugate rather than as an oligosaccharide. A hybrid glycoprotein that is prepared by using the NH_2 or COOH group of a glycoprotein will be as useful for this purpose as neoPG and pseudoPG probes. The preparation of probes and procedures for lectin screening are easily applied to probing glycoconjugates to search for biological ligands. We are currently developing a new method of probing glycoconjugates and will use it for screening receptors and characterizing lectins that have specificities useful for studying pathological states.

Procedures

Biotin-neoPG Probe Preparation

1. Dissolve GAG (3 mg) in 0.4 ml of distilled water in a 1.5 ml plastic tube.
2. Dissolve N-ethoxycarbonyl-2-ethoxy-1,2-dihydroquinoline (3 mg) in 0.6 ml of ethanol and add it to the tube.
3. Incubate the mixture at room temperature for 2 h with shaking.
4. Add BSA or HRP solution (6 mg/0.5 ml of distilled water) and incubate at 4°C for 2 days with gentle shaking.
5. Check the formation of neoPG by observing the change in the migration position of BSA or HRP on sodium dodecyl sulfate-polyacrylamide gel electrophoresis (SDS-PAGE)*.
6. For biotinylation of BSA-neoPG, add 222 μl of 0.1% N-hydroxysuccinimidylbiotin (NHS-biotin) to 3 ml of the sample in 50 mM sodium bicarbonate buffer (pH 8.5) and stand for 2 h in ice.
7. Ultrafiltrate the reaction mixture with Tris-buffered saline, pH 7.5, to remove excess reagent by using a membrane of exclusion molecular weight 50,000 or 100,000.
8. Choose the suitable concentration of the neoPG probe with a spot test and color development**.

Comments

*The coupling of GAG with protein is detected by staining the electrotransferred protein band on a PVDF membrane after SDS-PAGE with 0.5% toluidine blue and destaining with water.

**We typically use 10 μg/ml as a soluble probe for ELISA.

References

Chai W, Stoll MS, Galustian C, Lawson AM, Feizi T (2003) Neoglycolipid technology: deciphering information content of glycome. Methods Enzymol 362:160–195

Matsumoto H, Natsume A, Ueda H, Saito T, Ogawa H (2001) Screening of a unique lectin from 16 cultivable mushrooms with hybrid glycoprotein and neoproteoglycan probes and purification of a novel *N*-acetylglucosamine-specific lectin from *Oudemansiella platyphylla* fruiting body. Biochim Biophys Acta 1526:37–43

Ogawa H, Ueda H, Natsume A, and Suzuki R (2003) Preparation and utility of neoproteoglycan probes in the interaction analyses with glycosaminoglycan-binding proteins. Methods Enzymol 362:196–209

Satoh K, Yamagami M, Ogawa H (2004) Utility of pseudoproteoglycan (pseudoPG) probes that simulate PG supramolecular structure. Glycobiology 14:1095–1096

Ueda H, Kojima K, Saitoh T, Ogawa H (1999) Multispecificities of a *Psathyrella velutina* mushroom lectin: binding toward heparin/pectin is exhibited by a site different from its *N*-acetylglucosamine/*N*-acetylneuraminic acid-specific site. J Biochem Tokyo 126:530–537

Part 2
Chemical Synthesis of Sugar Chains

Section VI
Chemical Synthesis of Sugar Chains

Synthesis of Glycolipid and Its Application

Hideharu Ishida, Makoto Kiso

Introduction

Glycolipids constitute a group of bio-active glycoconjugates along with glycoproteins and proteoglycans. The synthesis of glycolipids by chemical methods is the most reliable procedure because of microheterogeniety carried mainly by lipid moiety, demonstrating good contrast to glycoproteins which are focused on homogeneity, now being synthesized by solid-phase, automated procedure in part.

In addition, design of useful analogs and derivatives of glycolipids to probe the structure–function relationship of these molecules is also a significant subject. Here, we describe the current and future scope on the synthesis of glycolipids, particularly glyco-sphingolipids, with regard to (1) constructions of oligosaccharides, (2) preparation of ceramides and the condensation with oligosaccharides, and (3) design of molecular probes based on glycolipids.

Construction of Oligosaccharides

Glycolipids are divided into groups such as sphingoglycolipids, glyceroglycolipids, and others including GPI and lipid A of endotoxin. Among them, sphingoglycolipids, particu-larly gangliosides, have attracted much attention because of their various biological activities.

Gangliosides are distinguished from other sphingoglycolipids, which contain sialic acid, and the chemical synthesis of gangliosides is attractive not only for biomedical applications but also for a challenging task in the field of synthetic chemistry.

Gangliosides are further divided into subclasses, such as ganglio-, lacto- and neolacto-, and globo- and isoglobo-series. The chemistry behind the construction of oligosaccharide includes two objectives: (1) construction of oligosaccharide backbones unique to each subclasses and (2) introduction of sialic acid into the backbone as a common moiety. The methodology has been rather matured, and a variety of gangliosides have been chemically synthesized (Ishida and Kiso 2001), including ganglioside GQ1b, which is one of the most complex ganglioside.

However, the methodology to scale up the materials in a large quantity, sufficient for biomedical applications has not yet been established. Recent advancements in the car-bohydrate chemistry including an efficient synthesis of α-sialosides (Ando and Imamura 2004) open the possibility that gangliosides can be supplied in several hundred milligrams.

Department of Applied Bio-organic Chemistry, Gifu University, Yanagido 1-1,
Gifu 501-1193, Japan
Phone: +81-58-293-2918, Fax: +81-58-293-2918
E-mail: ishida@gifu-u.ac.jp

Preparation of Ceramides and Condensation with Oligosaccharides

One of the reasons which makes the synthesis of gangliosides more difficult than those of glycopeptide, is the presence of ceramides. The problems concerning ceramides are (1) the preparation of ceramides and (2) the coupling to oligosaccharides. So far, sphingosine, a precursor and an equivalent of ceramides, have been synthesized in a diastereoselective manner from D-galactose, D-xylose, and L-serine as chiral sources. However, the methods are not even at a practical stage to supply sphingosine in an adequate amount. Recently, phytosphingosine, which has an additional hydroxyl group in place of double bond as shown in Fig. 1, has become commercially available in a large quantity at a reasonable price earlier manufactured by the fermentation of yeast, and the transformation of which into the sphingosine has been reported (Van den Berg et al. 2004).

Recently, biologically novel potentsphingoglycolipids have been isolated from organisms other than mammalians. One example is α-GalCer which has been isolated from marine sponge and found to activate Natural Killer T cell (NKT) (Kawano et al. 1997), and another is HLG-2 from sea cucumber, which activates the outgrowth of neurite as strong as the mammalian ganglioside GM3.

Both of them contain novel ceramides consisting of phytosphingosine, which motivates the significance of the distinct structures of ceramides for the biological activities. It is therefore crucial that sophisticated chemical synthesis of a variety of ceramides will be a guide to this area of bio-systems.

Furthermore, there is room for improvement in the introduction of ceramides to oligosaccharide. Intramolecular-aglycon delivery (IAD) and/or cassette synthesis of glycolipids employing glucosyl ceramide as a common building block can be one of the promising novel strategy to solve this issue, because it can supply ceramides in sufficient quantity.

Design of Molecular Probes Based on Glycolipids

In addition to the precise chemical synthesis of the natural-type complex glycolipids, the design and synthesis of their novel analogs and derivatives in which lipid moiety is simplified or removed is of great significance to probe the role of lipid part for the manifestation of biological activities. For example, we succeeded in synthesizing the useful analog of sialyl Lewis X ganglioside, namely GSC-150 (Hasegawa et al. 1995), based on the detailed structure–function relationship on the selectin–ligand interaction.

Ceramide was successfully simplified to a simple α-branchedalkanol (2-tetradecylhexadecanol) with the retention of activity as a ligand of selectin in this study. This designed ceramide is shown to replace the ceramides of sulfatide and sialylparagloboside (SGPG)also.

Oligosaccharides can exhibit their biological activities mainly by interacting with receptor proteins, and X-ray analysis is one of the most effective tool to elucidate the sugar–receptor interactions. It has been shown that the lipid moiety has tendency to prevent the formation of co-crystals of glycolipid and receptor proteins. Based on these findings, novel analogs which lack the lipid moiety and tethered with short tail such as 2-trimethylsilylethyl were designed to help the formation of co-crystals. In fact, 2-trimethylsilyl ethyl glycosides of disialyl Lewis a and ganglioside GT1a oligosaccharides made their co-crystals with Siglec-7 and tetanus toxin, respectively (Fotinou et al. 2001).

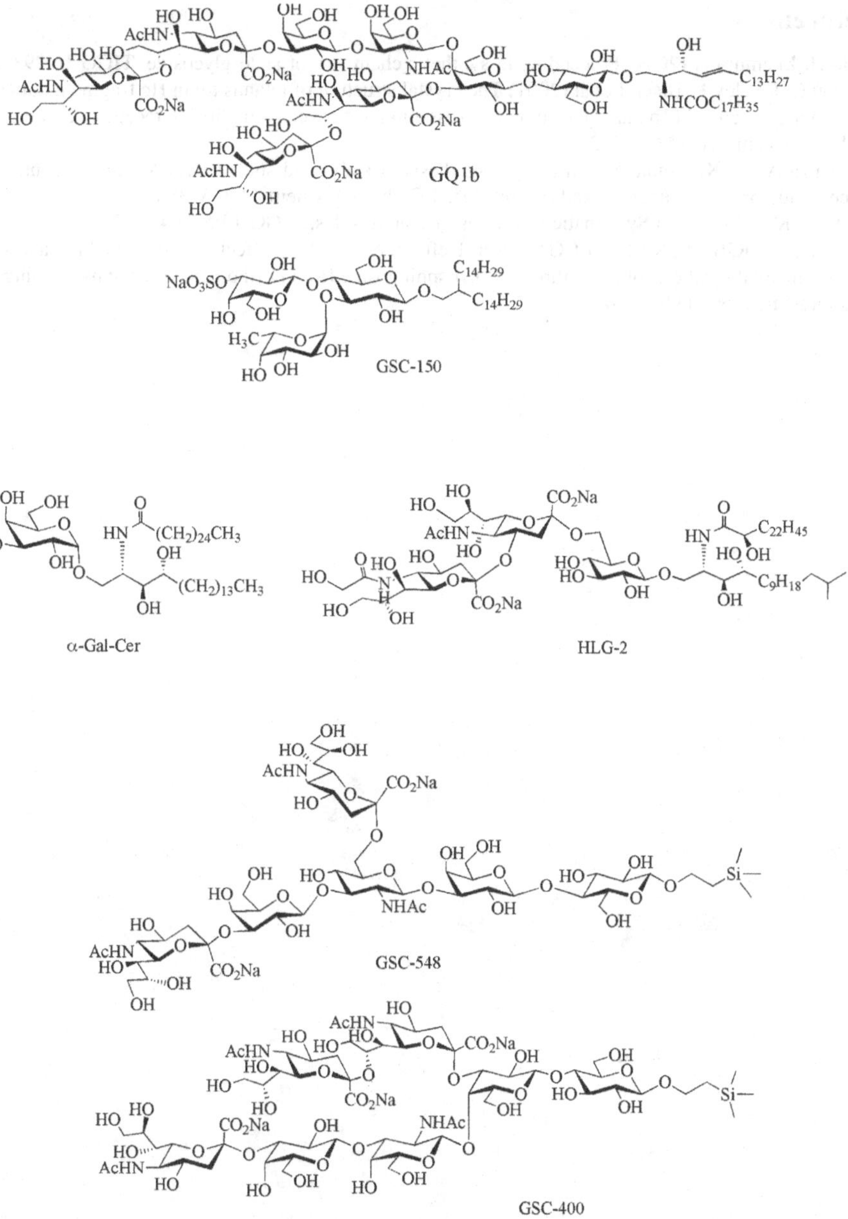

Fig. 1 Structures of synthetic glycans and glycolipids

As mentioned above, chemistry can contribute to the synthesis of glycolipids more substantially than that of glycoproteins. Although some problems remain to be unraveled, glycolipids have potent applications particularly in the biomedical field, as previously proved in the case of α-GalCer. In addition, it is of worth to note that glycolipids can exhibit their biological activities not only by acting solely but also by associating with membrane rafts and other structural elements, indicating that glycolipid may have interesting important roles in supramolecular chemistry.

References

Ando H, Imamura A (2004) Proceeding in synthetic chemistry of sialo-glycoside. TIGG 16:293–303

Fotinou C, Emsley P, Black I et al (2001) The crystal structure of tetanus toxin Hc fragment complexed with a synthetic GT1b analogue suggests cross-linking between ganglioside receptors and the toxin. J Biol Chem 276:32274–32281

Hasegawa A, Ito K, Ishida H et al (1995) Synthesis of sialyl and sulfo Lewis X analogs containing a ceramide or 2-(tetradecyl)hexadecyl residue. J Carbohydr Chem 14:353–368

Ishida H, Kiso M (2001) Systematic syntheses of gangliosides. TIGG 13:57–64

Van den Berg RJBHN, Korevaar CGN, Overkleeft HS et al (2004) Effective, high-yielding, and stereo-specific total synthesis of D-erythro-(2R,3S)-sphingosine from D-ribo-(2S,3S,4R)-phytosphingosine. J Org Chem 69:5699–5704

Chemo-Enzymatic Synthesis of Glycoconjugates

Toshiyuki Inazu

Introduction

Synthesis of glycoconjugate is one of the most important difficulties in carbohydrate chemistry. Although many chemical synthetic methods have been developed, the synthesis of glycoconjugates with natural glycan is difficult. Chemo-enzymatic syntheses of glycoconjugates using several kinds of enzymes have been reported. Sialic acid transferase has been used for the synthesis of sialo glycopeptides (Tanaka et al. 2006). Glycosylated amino acid derivative prepared using digestion of the peptide bond of glycoprotein by protease has also been applied to the synthesis of sialo glycopeptides (Yamamoto et al. 2003). In addition, the transglycosylation reaction by endo-β-N-acetylglucosaminidase (EC 3.2.1.96) has proved to be a useful method for glycoconjugate synthesis (Scheme 1).

Transglycosylation Reaction (Haneda et al. 2003)

Endo-β-N-acetylglucosaminidase hydrolyzes the N, N'-diacetylchitobiose moiety in N-glycoprotein. The hydrolysis products are the corresponding protein with N-acetyl-D-glucosamine (GlcNAc) residue and the oligosaccharide block lacking GlcNAc residue at the reducing end of N-glycan. Several microbial endo-β-N-acetylglucosaminidases, such as those derived from *Flavobacterium meningosepticum* (Endo-F), *Arthrobacter protophormiae* (Endo-A), and *Mucor hiemalis* (Endo-M), show transglycosylation activity. These enzymes transfer the oligosaccharide block to the suitable glycosyl acceptor, which has only one GlcNAc residue, to produce a glycoconjugate having natural N-glycan. Endo-A acts on a high mannose-type oligosaccharide, whereas Endo-M acts on complex-type, high mannose-type, and hybrid-type N-glycans. Recently, both native Endo-M of *Mucor hiemalis* and a recombinant Endo-M (TCI, Tokyo, Japan) are available. The recombinant Endo-M is more suitable, because this enzyme is free of protease activity and has the property of resistance to organic solvent. As a glycosyl donor for a disialo biantennary complex-type oligosaccharide, a sialylglycohexapeptide (SGP), H-Lys-Val-Ala-Asn[(NeuAc-Gal-GlcNAc-Man)$_2$-Man-GlcNAc$_2$]-Lys-Thr-OH, prepared from hen's egg yolk is also available (TCI, Tokyo & Otsuka Chemicals, Tokyo, Japan). The transglycosylated product is gradually hydrolyzed by Endo-M during longer incubation; therefore, the reaction must be terminated at the appropriate time.

Department of Applied Chemistry, School of Engineering, and Institute of Glycotechnology, Tokai University, 1117 Kitakaname, Hiratsuka, Kanagawa 259-1292, Japan
Phone: +81-463-58-1211, Fax: +81-463-50-2012
E-mail: inz@keyaki.cc.u-tokai.ac.jp

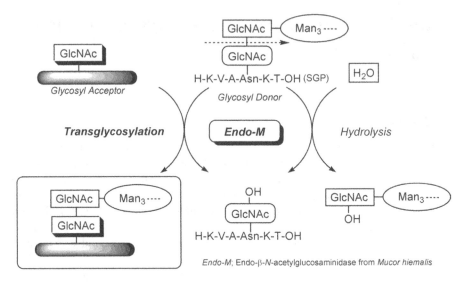

Scheme 1 Transglycosylation reaction using Endo-M

Synthesis of Glycopeptide with Natural N-Glycan (Mizuno et al. 1999)

We have developed a chemo-enzymatic synthetic method for glycopeptides combined with chemical synthesis of a glycopeptide containing GlcNAc as a glycosylation tag, and the transglycosylation catalyzed by Endo-M. This methodology includes the following four steps: (1) synthesis of an *N*-protected Asn(GlcNAc) derivative; (2) chemical synthesis of glycopeptides with one GlcNAc residue; (3) peptide elongation by the thioester segment condensation method; (4) enzymatic transglycosylation reaction of *N*-glycan by Endo-M. We have reported the synthesis of eel calcitonin (32 amino acid residues) derivatives with *N*-linked oligosaccharide on the Asn[3] residue as shown in Scheme 2.

Synthesis of Neoglycoconjugates (Haneda et al. 2003)

Endo-M strictly recognizes the equatorial 4-OH function of D-glycopyranose moiety of the acceptor substrate (GlcNAc, Glc, Man) and transglycosylates an oligosaccharide block onto it. On the other hand, the recognition specificity of Endo-M for the amino acid residue bound to GlcNAc is less stringent. Therefore, we synthesized several neoglycoconjugates using Endo-M. Substance P, H-Arg-Pro-Lys-Pro-Gln-Gln-Phe-Phe-Gly-Leu-Met-NH$_2$, is one of the bioactive peptides which has no Asn residue. We prepared GlcNAc-peptide using Fmoc-Gln(GlcNAc)-OH, which was synthesized by a method similar to that of Fmoc-Asn(GlcNAc)-OH. After enzymatic transglycosylation by Endo-M, we obtained the artificial derivative of Substance P with natural *N*-glycan on the Gln residue (Haneda et al. 2001). We also synthesized glycopeptides with multiple *N*-glycans. Cyclodextrin is a unique molecule, which possesses an aromatic compound in its cavity. Further, we applied to the synthesis of cyclodextrin analogs having *N*-glycan. We synthesized this cyclodextrin analog for the new drug delivery carrier.

Boc-Gly-SCH$_2$CH$_2$CO-Nle-NH-resin

$\downarrow$ ABI 430A Peptide Synthesizer
System Software Ver. 1.40 NMP/ HOBt t-Boc
End Capping by Ac$_2$O

H-Leu-Ser(Bzl)-Thr(Bzl)-Cys(Acm)-Val-Leu-
Gly-SCH$_2$CH$_2$CO-Nle-NH-resin

$\downarrow$ 1) Boc-Asn(**GlcNAc**)-O-P(S)Me$_2$
2) Boc-Ser(Bzl)-O-P(S)Me$_2$ 3) Fmoc-Cys(Acm)-O-P(S)Me$_2$

Fmoc-Cys(Acm)-Ser(Bzl)-Asn(**GlcNAc**)-Leu-Ser(Bzl)-Thr(Bzl)-
Cys(Acm)-Val-Leu-Gly-SCH$_2$CH$_2$CO-Nle-NH-resin

$\downarrow$ 1)10% Anisole/ HF 0°C 90 min
2) RP-HPLC

Fmoc-Cys(Acm)-Ser-Asn(**GlcNAc**)-Leu-Ser-   MALDI TOF-MS 1763.87
Thr-Cys(Acm)-Val-Leu-Gly-SCH$_2$CH$_2$CO-Nle-NH$_2$ Calcd 1764.0

$\downarrow$ 1) H-Lys(Boc)-Leu-Ser-Gln-Glu-Leu-His-Lys(Boc)-Leu-Gln-
Thr-Tyr-Pro-Arg-Thr-Asp-Val-Gly-Ala-Gly-Thr-Pro-NH$_2$
AgNO$_3$, DIEA, HOOBt/ DMSO
2) RP-HPLC

Fmoc-Cys(Acm)-Ser-Asn(**GlcNAc**)-Leu-Ser-Thr-Cys(Acm)-
Val-Leu-Gly-Lys(Boc)-Leu-Ser-Gln-Glu-Leu-His-Lys(Boc)-
Leu-Gln-Thr-Tyr-Pro-Arg-Thr-Asp-Val-Gly-Ala-Gly-Thr-Pro-NH$_2$

$\downarrow$ 1) 5% 1,4-Butandithiol/ TFA   MALDI TOF-MS 4183.82
2) 5% Piperidine/ DMSO   Calcd 4184.73

H-Cys(Acm)-Ser-Asn(**GlcNAc**)-Leu-Ser-Thr-Cys(Acm)-
Val-Leu-Gly-Lys-Leu-Ser-Gln-Glu-Leu-His-Lys-Leu-Gln-
Thr-Tyr-Pro-Arg-Thr-Asp-Val-Gly-Ala-Gly-Thr-Pro-NH$_2$

$\downarrow$ 1) AgNO$_3$, H$_2$O, DIEA/ DMSO
2) 1N HCl-DMSO
3) RP-HPLC

H-Cys-Ser-Asn(**GlcNAc**)-Leu-Ser-Thr-Cys-Val-Leu-
Gly-Lys-Leu-Ser-Gln-Glu-Leu-His-Lys-Leu-Gln-Thr-   MALDI TOF-MS 3619.02
Tyr-Pro-Arg-Thr-Asp-Val-Gly-Ala-Gly-Thr-Pro-NH$_2$ Calcd 3618.09

(Endo-M)

NeuAc-Gal-GlcNAc-Man$\searrow$
                                    Man-GlcNAc-GlcNAc
NeuAc-Gal-GlcNAc-Man$\nearrow$ |
                                             Asn

Glycosyl Donor: 25 mM
Glycosyl Acceptor: 10 mM   37°C
Endo-M: 4mU/ ml
Phosphate Buffer pH 6.25:60 mM   6 h

NeuAc-Gal-GlcNAc-Man$\searrow$
                                Man-GlcNAc-**GlcNAc**
NeuAc-Gal-GlcNAc-Man$\nearrow$ |
                              H-Cys-Ser-Asn-Leu-Ser-Thr-Cys-Val-Leu-

MALDI TOF-MS 5620.3   Gly-Lys-Leu-Ser-Gln-Glu-Leu-His-Lys-Leu-Gln-Thr-
    Calcd 5620.9   Tyr-Pro-Arg-Thr-Asp-Val-Gly-Ala-Gly-Thr-Pro-NH$_2$

Scheme 2 Chemo-enzymatic synthesis of [Asn(GlcNAc)[3]]-eel calcitonin

Protocol

Synthesis of Glycosylated eel Calcitonin (eCT)

[Asn(GlcNAc)^3]-eCT (22.1 mg, 6.1 µmol) and SGP (172 mg, 60 µmol) were dissolved with 360 µL of 0.1 M potassium phosphate buffer, pH 6.25, in a 2 mL-Eppendorf tube. After preincubation at 37°C for 5 min, 240 µL of Endo-M solution (final concentration, 60 mU/mL) was added. The reaction mixture (600 mL) was incubated at 37°C for 30 min with gentle stirring. The reaction was terminated by addition of an equal volume of 0.5% cold TFA solution. The transglycosylation product was purified by preparative RP-HPLC (ODS column, 20 × 250 mm), and freeze-dried. The transglycosylated product, [Asn{(NeuAc-Gal-GlcNAc-Man)$_2$-Man-GlcNAc$_2$}^{3}]-eCT (16.7 mg, 2.96 µmol), was obtained as a white powder in 49% yield.

References

Haneda K, Inazu T, Mizuno M, Iguchi R, Tanabe H, Fujimori K, Yamamoto K, Kumagai H, Tsumori K, Munekata E (2001) Chemo–enzymatic Synthesis of a Bioactive Peptide Containing a Glutamine–linked Oligosaccharide and Its Charactarization. Biochim Biophys Acta 1526:242

Haneda K, Inazu T, Mizuno M, Yamamoto K (2003) Chemo–enzymatic Synthesis of Neo–glycopeptides Using Endo–M. Methods Enzymol 362(A):74

Mizuno M, Haneda K, Iguchi R, Muramoto I, Kawakami T, Aimoto S, Yamamoto K, Inazu T (1999) Synthesis of Glycopeptide Having Oligosaccharide: Chemo–enzymatic Synthesis of Eel Calcitonin Analogs Having Natural N–Linked Oligosaccharides. J Am Chem Soc 121:284

Tanaka E, Nakahara Y, Kuroda Y, Takano Y, Kojima N, Hojo H, Nakahara Y(2006) Chemoenzymic synthesis of a MUC1 glycopeptide carrying non–natural sialyl TF–β O–glycan. Biosci Biotechnol Biochem 70:2525

Yamamoto N, Ohmori Y, Sakakibara T, Sasaki K, Juneja LR, Kajihara Y (2003) Solid–phase synthesis of sialylglycopeptides through selective esterification of the sialic acid residues of an Asn–linked complex–type sialyloligosaccharide. Angew Chem Int Ed 42:2537

Construction of *O*-Linked Glycopeptide Library Using Human Glycosyltransferases

Hiromi Ito

Introduction

Biological functions of glycoconjugates and biomedical applications of carbohydrates have gathered researcher's attention. The construction of oligosaccharide/glycopeptide libraries would be essential for the development of carbohydrate-based research. We introduce, here, a novel method for the construction of structurally defined glycopeptide library which can be easily converted to an oligosaccharide library using human recombinant glycosyltransferases. In the past few years, approximately 180 glycogenes, which encode glycosyltransferases, sulfotransferases, and sugar-nucleotide transporters, had been cloned. We had created many human glycosyltransferases as a recombinant enzyme, and consequently it is possible to prepare libraries which contain a large variety of oligosaccharide structures using these enzymes.

The advantages of our method named Enzyme-Cue Synthesis are as follows. First, the formation of glycosodic linkage can be easily controlled by glycosyltransferases which have an excellent regio- and stereo-selectivity. Second, by repeating incomplete enzyme reactions, i.e., by terminating each reaction halfway intentionally in a single tube, it is possible to obtain a diversity of oligosaccharides which have theoretically 2^n (n is the number of reaction steps) structures. Third, each structure of components in a library was designed to match with (or have) a certain molecular weight i.e., each product has a different m/z value. According to this strategy which involves a sequential addition of several glycosyltransferases, we succeeded in producing an *O*-linked glycopeptide library which is composed of different components in a single reaction tube. Since the components of the library obtained exhibited different m/z values, their glycan structures were easily determined by mass spectrometry alone.

Because the *O*-linked glycopeptide library obtained by this method consists of compounds which have structurally defined glycans, the glycopeptide library is (1) used as a standard material for MS^n spectral database in glycan structural analysis using multistage tandem mass spectrometry after conversion to a glycan library by reductive β-elimination, and (2) used as a tool to assay glycan-recognition specificity of lectins and antibody, etc.

Research Center for Medical Glycoscience, National Institute of Advanced Industrial Science and Technology (AIST), 1-1-1 Umezono, Tsukuba, Ibaraki 305-8568, Japan
E-mail: hiromi-itou@aist.go.jp

Procedure

I. Preparation of Human Recombinant Enzymes

Materials

Human embryonic kidney (HEK) 293T cells as expression system
pFLAG-CMV-1-ST3GalIV, pFLAG-CMV-1-FUT6, pFLAG-CMV-3-DEST-β3Gn-T2, and pFLAG-CMV-3-DEST-β4GalT-I as transfected plasmids
LipofectAMINE 2000 (Invitrogen)
Opti-MEM (Invitrogen)
ANTI-FLAG M1 Affinity Gel (Sigma-Aldrich)
50 mM Tris-buffered saline (50 mM Tris–HCl, pH 7.4 and 150 mM NaCl prepared in MilliQ water)

Methods

1. Glycosyltransferase genes are engineered for heterologous expression in mammalian cells as fusion proteins containing a FLAG Tag at the *N*-terminus.
2. The putative catalytic domain of each glycosyltransferase gene, without a transmembrane domain, is subcloned into a mammalian expression vector.
3. About 30 µg of each resulting plasmid is transfected into HEK293T cells (2×10^6) using LipofectAMINE 2000 according to the manufacturer's instructions.
4. After incubation at 37°C for 72 h, collect the culture medium.
5. Add ANTI-FLAG M1 affinity gel to 10–50 mL of each culture medium from HEK293T cells. Incubate at 4°C overnight, with gentle rotating.
6. Wash the resin 5 times with 50 mM Tris-buffered saline containing 1 mM CaCl$_2$ and suspended in 100 µL of the same buffer.

Example of the Preparation of O-Linked Glycopeptide Library

Materials

500 mM HEPES-NaOH buffer, pH 7.0
200 mM MnCl$_2$
Purified recombinant enzymes bound to anti-FLAG M1 affinity gel
An appropriate concentration of sugar nucleotides (CMP-Neu5Ac, UDP-GlcNAc, GDP-Fuc, and UDP-Gal; Sigma-Aldrich)
A carboxyfluorescein (FAM)-labeled Muc1a tandem repeat peptide

Methods

1. Prepare a reaction buffer in a final volume of 40 µL; 50 mM HEPES-NaOH buffer, pH 7.0, which contains 10 mM MnCl$_2$, 500 µM CMP-Neu5Ac, and 25 µM of starting glycopeptide (Galβ1-4GlcNAcβ1-6(Galβ1-3)GalNAc on the FAM-labeled MUC1a tandem repeat peptide).
2. Add the purified ST3GalIV to the reaction mixture. Incubate at 37°C for 20 h.
3. Inactivate ST3GalIV at 100°C for 5 min. Add 1 µL of 1 mM UDP-GlcNAc and the purified β3Gn-T2 to the reaction solution. Incubated at 37°C for 2 h.
4. Inactivate β3Gn-T2 at 100°C for 5 min. Add 1 µL of 1 mM GDP-Fuc and the purified FUT6 to the reaction mixture. Incubate the mixture at 37°C for 30 min.

5. Inactivate FUT6 at 100°C for 5 min. Add 1 μL of 500 μM UDP-Gal and the purified β4GalT-I to the mixture. Incubate the mixture at 25°C for 2 h.
6. Inactivate β4GalT-I at 100°C for 5 min. Remove enzymes with resin using an Ultrafree-MC column (Millipore).
7. Purify glycopeptides with a reversed-phased microcolumn, ZipTip C18 (Millipore) or Oasis HLB (10 mg/1 cc; Waters).

References

Ito H, Kameyama A, Sato T, Kiyohara K, Nakahara Y, Narimatsu H (2005) Molecular-weight-tagged glycopeptide library: efficient construction and applications. Angew Chem Int Ed 44:4547–4549

Kameyama A, Kikuchi N, Nakaya S, Ito H, Sato T, Shikanai T, Takahashi Y, Takahashi K, Narimatsu H (2005) A Strategy for identification of oligosaccharide structures using observational multistage mass spectral library. Anal Chem 77:4719–4725

Narimatsu H (2004) Construction of a human glycogene library and comprehensive functional analysis. Glycoconj J 21:17–24

Sato T, Gotoh M, Kiyohara K, Kameyama A, Kubota T, Kikuchi N, Ishizuka Y, Iwasaki H, Togayachi A, Kudo T, Ohkura T, Nakanishi H, Narimatsu H (2003) Molecular cloning and characterization of a novel human β1,3-glucosyltransferase, β3Glc-T, which is localized at the endoplasmic reticulum and glucosylates O-linked fucosylglycan on thrombospondin type 1 repeat domain. J Biol Chem 278:47534–47544

Taniguchi N, Honke K, Fukuda M (eds) (2001) Handbook of glycosyltransferases and related genes. Springer, Tokyo

Alteration of *N*-Linked Oligosaccharides

Yasuhiro Kajihara

Introduction

Oligosaccharides are linked to the protein surface and play roles in a number of biological events. Therefore, much attention is being paid to research and investigate the function of the oligosaccharides. Oligosaccharides on glycoprotein are roughly divided into two groups: *O*-linked and *N*-linked types. The *N*-linked type is a large branched oligosaccharide, which is further divided into three types: complex, hybrid, and high-mannose type. All oligosaccharides are linked to the nitrogen of asparagine by an *N*-glycosyl linkage. In order to prove the function of *N*-glycan on the protein surface, a convenient synthetic method for *N*-glycan chain should be developed. However, the chemical synthesis is time-consuming due to repetitive protection/deprotection steps. *N*-Glycans in glycoprotein show structural diversity in its oligosaccharides, the so-called glycoform. Here, I would like to introduce recent synthetic developments focusing on diverse *N*-linked oligosaccharides based on semisynthesis.

We used asparagine-linked biantennary complex-type-sialylundecasaccharide **1** (Seko et al. 1997) obtained from egg yolk to prepare more than 20 kinds of a pure *N*-linked oligosaccharide (Asn-oligosaccharide) using branch-specific glycosidase digestion (Kajihara et al. 2004). In short mono sialyloligosaccharides was prepared from **2** by acid hydrolysis of NeuAc, and subsequent exo-glycosidase digestion (β-galactosidase, *N*-acetyl-β-D-glucosaminidase and α-D-mannosidase) of the individual asialo-branch afforded corresponding diverse oligosaccharides. Asn-oligosaccharide **2** was prepared by use of protease (Actinase-E) digestion from **1**. To release one of the two NeuAc residues from **2**, acid hydrolysis of Asn-oligosaccharide **2** by 40 mM HCl solution was performed. This reaction afforded 4 kinds of Asn-oligosaccharide **2**, **3**, **4**, and **5** along with **6** and **7** as contaminants. In order to obtain pure mono-sialyloligosaccharide **3** and **4**, oligosaccharides **2–7** were protected by hydrophobic protecting group such as 9-fluorenylmethyl group (Fmoc) in order to increase their hydrophobicity (Scheme 2). As expected, this increased hydrophobicity of Fmoc-oligosaccharides **8–13** enabled us to purify each Asn-oligosaccharide by an ODS-column chromatography except for **9** and **10**. The mixture of **9** and **10** was further protected by forming benzyl ester of NeuAc. This treatment enabled us to purify each monosialyloligosaccharides **14** and **15** (Scheme 3).

Galactosidase digestion of **9** and **10** afforded nonasaccharides **12** and **16**. The limit of this purification is about 20 mg in one purification step by HPLC (ODS: ϕ 20 × 250 mm). Repetitive chromatography on the same column yielded 200 mg of each monosialyloli-

International Graduate School of Arts and Sciences, Yokohama City University, 22-2 Seto, Kanazawa-ku, Yokohama, Kanagawa 236-0027, Japan
Phone: +81-45-787-2210, Fax: +81-45-787-2413
E-mail: kajihara@yokohama-cu.ac.jp

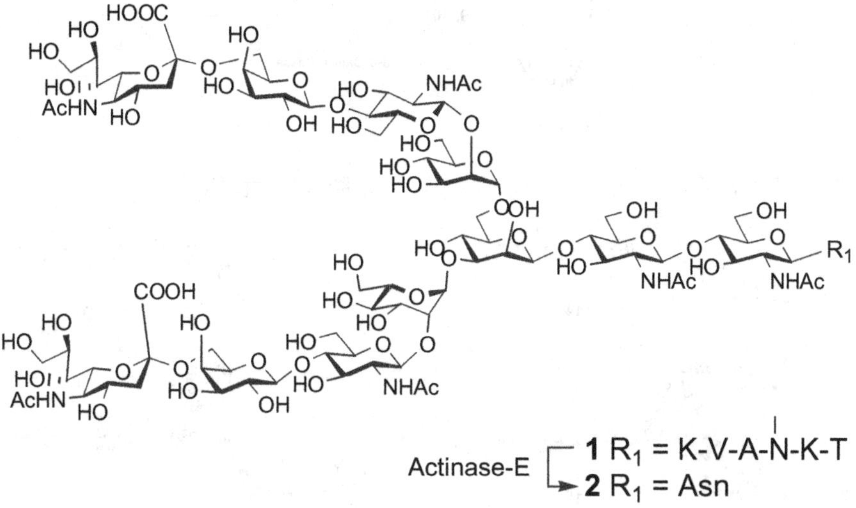

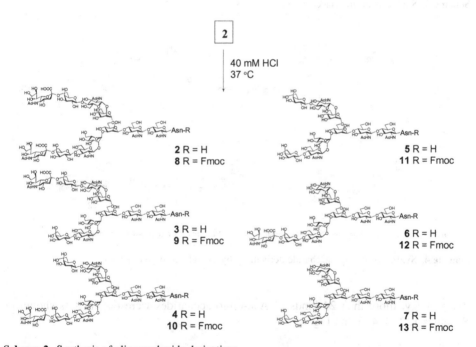

Actinase-E ⌈ **1** R_1 = K-V-A-N-K-T
            └→ **2** R_1 = Asn

Scheme 1 Synthesis of *N*-linked oligosaccharide

Scheme 2 Synthesis of oligosaccharide derivatives

gosaccharide **12** and **16**. Each monosialyloligosaccharide, **12** and **16**, was separately treated by exo-glycosidase digestion as shown in Schemes 4 and 5. Each exo-glycosidase digestion can be performed in 100 mg-scale and yield ranged from 70 to 90% (isolated yield). Asn-oligosaccharides shown in Fig. 1 can also be prepared from **12** by the same strategy shown in Schemes 4 and 5. The structure of these oligosaccharides thus obtained was determined by use of reporter group and high-resolution mass spectroscopy. Using

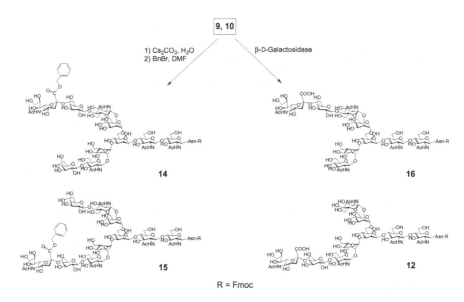

Scheme 3 Synthesis of oligosaccharide derivatives

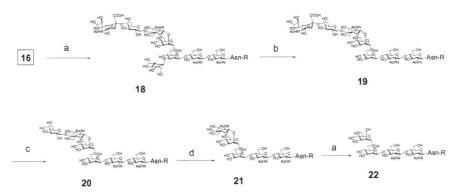

a) *N*-acetyl-β-D-Glucosaminidase b) α-D-Mannosidase c) α-D-Neuraminidase d) β-D-Galactosidase. R = Fmoc.

Scheme 4 Synthesis of oligosaccharide derivatives by exo-glycosidase digestion

this strategy, more than 24 kinds of Asn-oligosaccharides were obtained as shown in Schemes 1, 2, 3, 4, 5 and Fig. 1.

Procedures

Preparation 2

1. Dissolve sialylglycopeptide **1** (60 mg) and NaN_3 in a solution of a Tris–HCl buffer (50 mM, $CaCl_2$ 10 mM, pH 7.5, 3 mL).
2. Add Actinase-E (9 mg) in the buffer solution and incubate for 2 days at 37°C (during incubation, the pH was kept at 7.5).
3. Monitor this reaction by TLC (1M NH_4OAc : isopropanol = 1 : 1).

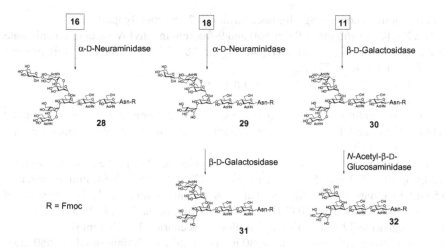

Scheme 5 Synthesis of oligosaccharide derivatives by exo-glycosidase digestion

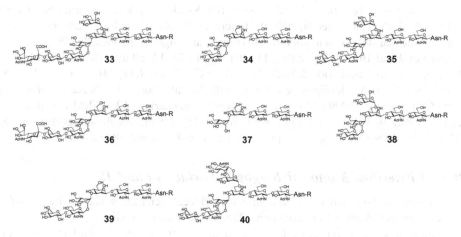

Fig. 1 Oligosaccharide library: structure of oligosaccharide derivates. R = Fmoc group

4. Lyophilize after finishing this reaction.
5. Purify the residue by gel permeation (Sephadex-G-25, φ 2.0 × 20 cm, H$_2$O) to afford pure *N*-linked sialyloligosaccharide **2** (42 mg, 86%).

Fmoc-Oligosaccharide 8, 9, 10, and 11

1. Add HCl solution (80 mM, 11.4 mL) to water solution (11.4 mL) containing Asn-oligosaccharide **2** (1.07 g, 456 μmol)
2. Incubate this solution for 6 h at 37°C.
3. Allow to cool to 4°C and neutralize with aq NaHCO$_3$.
4. Lyophilize this solution.
5. Purify this residue by gel permeation column (Sephadex G-25: φ 2.5 × 100 cm, water) to afford a mixture (778 mg) of disialo substrate **2**, monosialyl oligosaccharides **3**, **4** and asialooligosaccharide **5** along with **6** and **7**.

6. Pool fractions containing oligosaccharides **2–7** and then lyophilize.

7. Add $NaHCO_3$ (162 mg, 1.93 mmol) and 9-fluorenylmethyl-*N*-succimidylcarbonate (432 mg, 1.28 mmol) to a solution (H_2O-acetone; 3.8 –5.7 mL) containing this mixture (**2**, **3**, **4**, **5**, **6**, and **7**: total 778 mg).

8. Stir this solution for 2 h at room temperature.

9. Evaporate to remove acetone.

10. Inject this solution to HPLC system using ODS-column (ϕ 20 × 250 mm; eluted with H_2O, 100 mL and then 25% MeCN, 200 mL) to afford a mixture of Fmoc-oligosaccharides **8–13** (681 mg).

11. Purify this mixture by HPLC system (ODS column; YMC packed column D-ODS-5-A 120A, ϕ 20 × 250 mm, 50 mM NH_4OAc : MeCN = 82 : 18, 7.5 mL/min, monitoring at 215 nm) to obtain disialyloligosaccharide **8** (retention time 15.7 min), mixture of monosialyloligosaccharide **9**, **10** (19.5 min), asialooligosaccharide **11** (25.5 min), monosialylnonasaccharide **12** (23.3 min) and asialooctasaccharide **13** (28.2 min).

12. Repeat this purification with ca. 30 mg of mixture each time (total ca. 680 mg).

13. Combine the individual oligosaccharides thus obtained.

14. Lyophilize these oligosaccharide solutions.

15. Inject each residue to HPLC system (ODS-column: ϕ 5 × 150 mm; eluted with H_2O, 100 mL then 25% MeCN, 200 mL) to remove NH_4OAc salt. Pool fractions containing oligosaccharide and then lyophilize to obtain pure oligosaccharides. **8** (148 mg, 13%), mixture of **9** and **10** (249 mg, 24%), **11** (101 mg, 11%), **12** (68 mg, 7%) and **13** (35 mg, 4%). Di-sialooligosaccharide **8**: HRMS Calcd for $C_{103}H_{154}N_8NaO_{66}$ [$M + Na^+$] 2581.8838, found 2581.8821; Asialooligosaccharide **11**: HRMS Calcd for $C_{81}H_{120}N_6NaO_{50}$ [$M + Na^+$] 1999.6930, found 1999.6939; Mono-sialyloligosaccharide **12**: HRMS Calcd for $C_{86}H_{125}N_7Na_3O_{53}$ [$M + Na^+$] 2172.6995, found 2172.7084; Asialooctasaccharide **13**: HRMS Calcd for $C_{75}H_{110}N_6NaO_{45}$ [$M + Na^+$] 1837.6402, found 1837.6418.

Benzyl Esterified Monosialyloligosaccharides 14 and 15

1. Remove sodium salt from Fmoc-monosialyldecasaccharide **9** and **10** by use of a column (ϕ 0.5 cm × 5 cm containing Dowex-50Wx8(H^+) resin).

2. Apply a solution (H_2O : 1 mL, 4°C) containing **9** and **10** (5.0 mg) to the anion exchange column.

3. Wash the column by 10 mL of cold water.

4. Pool fractions containing **9** and **10** and then lyophilize.

5. Dissolve this residue in H_2O (0.22 mL).

6. Neutralize this solution to pH 7 by stepwise addition of a solution of Cs_2CO_3 (2.5 mg/ml) monitoring with pH meter.

7. Lyophilize the solution after neutralization.

8. Dissolve this residue in dry DMF (0.43 mL).

9. Add a solution of BnBr (6.6 μL) in DMF (20 μL) to the DMF solution containing **9** and **10**.

10. Stir this mixture at room temperature under argon atmosphere.

11. Add diethyl ether (5 mL) to this mixture after 48 h.

12. Collect precipitate formed by centrifugation.

13. Inject a solution containing this residue to HPLC system (ODS column, ϕ 20 × 250 mm, 50 mM NH_4OAc : MeCN = 78 : 22) to obtain mono-benzyl-sialyloligosaccharides **14** (91 min) and **15** (88 min).

14. Pool fractions containing each oligosaccharides and then concentrate in vacuo.
15. Inject solution containing each residue to HPLC system (ODS-column: ϕ 5 × 150 mm; H_2O 50 mL and then 25% MeCN, 100 mL) to remove NH_4OAc salt.
16. Pool fractions containing each oligosaccharide and lyophilize to obtain pure mono-benzyl-sialyloligosaccharides, **14** (1.6 mg) and **15** (1.8 mg). **14**: MS(Fab) Calcd for $C_{99}H_{143}N_{17}NaO_{58}$ $[M + H^+]$ 2380.8, found 2380.1; **15**: MS(Fab) Calcd for $C_{99}H_{143}N_7NaO_{58}$ $[M + H^+]$ 2380.8, found 2380.5.

General Procedure of Glycosidase Digestion (Example: Galactosidase Digestion of Mono-Sialyldecasaccharide 9 and 10)

1. Dissolve monosialyldecasaccharides **9** and **10** (135 mg, 59.4 μmol), bovine serum albumin (1.0 mg) and β-D-galactosidase (390 mU) in a HEPES buffer (50 mM, pH 6.0, 5.6 mL) solution.
2. Incubate this mixture at 37°C for 19 h, and then lyophilize.
3. Inject the residue to HPLC (YMC packed column D-ODS-5-A120A, ϕ 20 × 250 mm, 50 mM NH_4OAc : MeCN = 82 : 18, 7.5 mL/min) to obtain mono-sialyl oligosaccharide **16** (36 min) and **12** (39.2 min).
4. Repeat this purification with ca. 10–20 mg portions of the reaction mixture (total ca. 120 mg).
5. Combine the individual oligosaccharides thus obtained.
6. Lyophilize these oligosaccharides.
7. Inject sample to HPLC system (ODS-column: ϕ 5 × 150 mm; H_2O 100 mL and then 25% MeCN solution 200 mL) to obtain pure oligosaccharides **16** (51 mg, 41%) and **12** (60 mg, 48%). Mono-sialylnonasaccharide **16**: HRMS Calcd for $C_{86}H_{127}N_7NaO_{53}$ $[M + Na^+]$ 2128.7356, found 2128.7363.

References

Kajihara Y, Suzuki Y, Yamamoto N, Sasaki K, Sakakibara T, Juneja LR (2004) Prompt chemo-enzymatic synthesis of diverse complex-type oligosaccharides and its application to the solid-phase synthesis of a glycopeptide with Asn-linked sialyl-undeca- and asialo-nona-saccharides. Chem Eur J 10(4):971–985

Seko A, Koketsu M, Nishizono M, Enoki Y, Ibrahim HR, Juneja LR, Kim M, Yamamoto T (1997) Occurrence of a sialylglycopeptide and free sialylglycans in Hen's egg yolk. Biochim Biophys Acta 1335:23

Stationary Solid-Phase Reaction (SSPR) for Oligosaccharide Synthesis

Takuro Ako, Osamu Kanie

Introduction

Synthetic methods to access oligosaccharide structures are extremely important to provide bio-probes for the functional analysis. Furthermore, a combinatorial oligosaccharide library is potentially important in terms of diagnoses and providing potential seeds for new pharmaceuticals.

Solid-phase synthesis of oligosaccharides has been a focus area because it relates to a rapid access to such molecules and important achievements have been reported. One of the goals of this trend of researches is the establishment of a solid-phase automated oligosaccharide synthesizer.

Despite the realization and needs of the method, it has been difficult to accomplish saccharide synthesis in high yield especially one starts the synthesis form non-reducing end monosaccharide. Introduction of a stationary solid-phase reaction, where any mechanical stirring method is applied, enabled high concentration condition leading to high yielding reaction (Ako et al. 2006; Kanie et al. 2006).

Procedure

Typical experimental method for the synthesis of **Resin-bound 2,3-di-O-benzyl-α,β-L-fucopyranosyl-(1→6)-2,3,4-tri-O-benzyl-α,β-D-galactopyranosyl fluoride (3)** (Fig. 1). Methyl trifluoromethanesulfonate (MeOTf, 11.6 μL, 101 μmol) was added to dimethyl disulfide (MeSSMe, 9.2 μL, 101 μmol) and stirred at room temperature for 5 min, and then the mixture was diluted with dichloroethane (300 μL) and CH_3CN (300 μL) to give 0.168 M DMTST (MeSS$^+$Me$_2$.$^-$OTf) stock solution. This solution (530 μL, 89.0 μmol) was mixed with an acceptor, 2,3,4-tri-O-benzyl-α,β-D-galactopyranosyl fluoride (**2**, 26.6 mg, 58.8 μmol, $\alpha : \beta = 21 : 1$) at 15°C to give a 0.111 M acceptor/0.168 M DMTST solution. Compound **1** (100.4 mg, 27.9 μmol) was swollen by addition of the DMTST-acceptor solution (500 μL, acceptor: 55.5 μmol, DMTST: 84.0 μmol) at 15°C and the reaction was allowed to become 0°C with temperature increment rate of 1.25°C/h. The reaction was carried out without any mechanical stirring nor flow. After completion of the reaction, the resin was washed successively with dichloromethane, DMF, and dichloromethane to afford **3** (Fig. 2).

Mitsubishi Kagaku Institute of Life Sciences (MITILS), Minamiooya 11, Machida, Tokyo 194-8511, Japan
Phone: +81-42-724-6238, Fax: +81-42-724-6317
E-mail: kokanee@mitils.jp

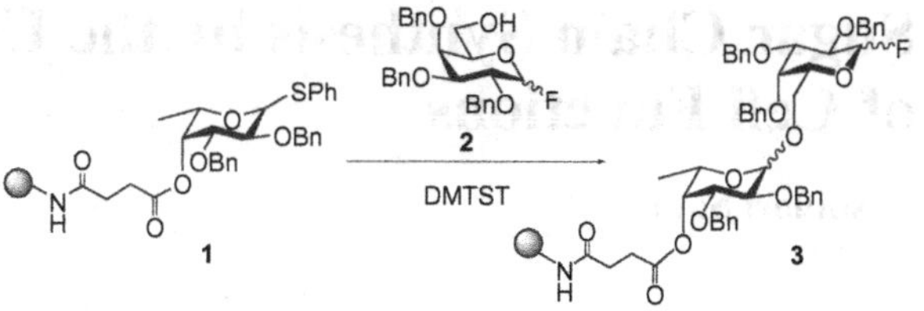

Fig. 1 Stationary solid-phase reaction of fucosyl donor (phenylthio glycoside) and galactosyl acceptor (glycosyl fluoride)

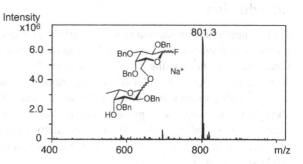

Fig. 2 Mass spectrum of a formed disaccharide just after cleavage of the product from the resin using NaOMe in dichloromethane-MeOH at room temperature

References

Ako T, Daikoku S, Ohtsuka I, Kato R, Kanie O (2006) A method of orthogonal oligosaccharide synthesis leading to combinatorial library based on stationary solid-phase reaction. Chem Asian J 1:798–813

Kanie O, Ohtsuka I, Ako T, Daikoku S, Kanie Y, Kato R (2006) Orthogonal glycosylation reactions on solid-phase and the synthesis of a library consisting of a complete set of fucosyl galactose isomers. Angew Chem Int Ed 45:3851–3854

Sugar Chain Synthesis by the Use of Cell Functions

Toshinori Sato

Introduction

For the supply of oligosaccharides as materials, extraction from natural resources, organic synthesis and enzymatic synthesis have been carried out. We developed a method of sugar chain synthesis using cell function as the fourth method (Sato et al. 2007) (Fig. 1). As the cells express a number of glycosyltransferase depending on the cell lines, which are the small factories synthesizing the sugar chain. By administering the amphiphilic alkylglycosides (saccharide primers) into the cell culture medium, the saccharide primers were glyco-sylated by cells. By biocombinatorial synthesis, i.e., combination of the saccharide primers and established cell lines, a number of oligosaccharides have been synthesized (Fig. 1).

Saccharide Primer Method

To synthesize oligosaccharides using cell function, saccharide primers that are substrates for glycosyltransferase in cells were employed. The first saccharide primer, dodecyl β-lactoside (Lac-C12), is a mimic of lactosylceramide (LacCer), which is a common precursor for glycolipids including ganglio series, globo series, lacto series, and neo-lacto series (Miura et al. 1997). Amphiphilic Lac-C12 could internalize into cells when supplemented in the culture medium. Most of the products glycosylated in the cells were secreted in the culture medium. Since the glycosylated products existed in the culture medium, the isolation was very convenient. This is the most important advantage of the method of synthesizing oligosaccharides using cells. By utilizing the intact cells, it is not necessary to prepare glycosyltransferase and glycosyl donor.

Experimental procedures were as follows: the cells were cultured in the presence of the saccharide primer in a serum-free medium for 1–2 days. The glycosylated products in the culture medium were collected using a reversed-phase column chromatography, and were separated by HPTLC or HPLC. Their molecular masses and sequences were analyzed by MALDI-TOF–MS and ESI–MS.

Biocombinatorial Synthesis

The principle of biocombinatorial synthesis was employed for the preparation of oligosaccharide library. Each established cell line has specific glycan biosynthesis pathways. Lac-C12 was elongated with sugar chain as equal to the endogenous glycolipids such as

Department of Biosciences and Informatics, Keio University, Hiyoshi, Yokohama 223-8522, Japan
Phone: +81-45-566-1771, Fax: +81-45-566-1447
E-mail: sato@bio.keio.ac.jp

Saccharide Primer Method   ## Biocombinatorial Synthesis

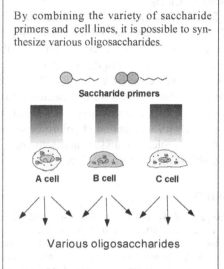

Fig. 1 The principle of saccharide primer method and biocombinatorial synthesis

ganglio series, globo series, lacto series, or neo-lacto series. In addition to Lac-C12, the several saccharide primers having monosaccharides (*N*-acetylglucosamine (Sato et al. 2008) or galactose), disaccharides, and a sugar–amino acid conjugate were also synthesized, and were found to be glycosylated in the cells. By combining the culture cells and the saccharide primers, it is possible to produce many kinds of oligosaccharides that are produced by the cells.

The saccharide primer, 12-azidododecyl glycoside (Lac-C12-N$_3$), having reaction active group at the end of alkyl chain was also glycosylated in the cells (Kasuya et al. 2000). The azido glycosides were successfully immobilized to a sensor chip or ELISA plate by either the Staudinger reaction or reduction of azido group followed by condensation reaction (Sato et al. 2004). Specific bindings of lectin or antibody to the immobilized glycan were achieved by both the methods. Therefore, the saccharide primer method would be useful for the development of glycan array.

Saccharide Primer Method

Saccharide primers (alkyl glycosides), that are substrates for glycosyltransferase in cells, were employed to synthesize oligosaccharides using cell function.

1. The cells (2×10^6 cells) were cultured in a serum-free medium supplemented with a saccharide primer (50 µM) for 1–2 days.

2. The unreacted primer and glycosylated products in the culture medium were collected using a reversed-phase column chromatography (Sep-Pak-C18).

3. The unreacted primer and the glycosylated products were eluted with methanol/chloroform-mixed solvents.

4. The eluate containing the glycosylated products was evaporated under reduced pressure. The obtained products were dissolved in chloroform/methanol (2:1),

and an aliquot was separated on a HPTLC plate (Silica gel 60) using chloroform/methanol/0.2% $CaCl_2$.

5. TLC blotting was carried out as follows: glycosylated products separated on a HPTLC plate were sprayed with primuline reagent and the spots were marked with a red pencil under UV light. Then, the HPTLC plate was dipped in a blotting solvent containing 2-propanol/methanol/0.2% $CaCl_2$ (40 : 7 : 20) for 20 s and placed on a glass fiber filter. The plate was covered with a PVDF membrane, a PTFE membrane and another glass fiber filter. These layers were subjected to pressure at 180°C for 30 s using a TLC thermal blotter. The PVDF membrane was washed with pure water, and glycolipid fractions were extracted with methanol and chloroform/methanol (2 : 1, by volume).

6. The molecular masses and sequences of the products were analyzed by MALDI–TOF–MS/MS and ESI–MS/MS.

References

Kasuya MCZ, Wang LX, Lee YC, Mitsuki M, Nakajima H, Miura Y, Sato T, Hatanaka K, Yamagata S, Yamagata T (2000) Azido glycoside primer: a versatile building block for the biocombinatorial synthesis of glycosphingolipid analogues. Carbohydr Res 329:755–763

Miura Y, Yamagata T (1997) Glycosylation of lactosylceramide analogs in animal cells: amphipathic disaccharide primers for glycosphingolipid synthesis. Biochem Biophys Res Commun 241:698–703

Sato T, Fujita S, Kasuya MCZ, Hatanaka K, Yamagata T (2004) Display of azido glycoside on a sensor chip. Chem Lett 33:580–581

Sato T, Hatanaka K, Hashimoto H, Yamagata T (2007) Syntheses of oligosaccharides using cell function. TIGG 19:1–17

Sato T, Takashiba M, Hayashi R, Zhu X, Yamagata T (2008) Glycosylation of dodecyl 2-acetamide-2-deoxy-β-D-glucopyranoside and dodecyl β-D-galactopyranosyl-(1→4)-2-acetamide-2-deoxy-β-D-glucopyranoside as saccharide primers. Carbohydr Res 343:831–838

Glyco-Chemistry Cycle System Based on Glycosidases

Shin-ichiro Shoda

Introduction

Carbohydrates are photosynthesized using carbon dioxide and water and are the most widely distributed organic compounds as biomass on earth. Almost all naturally occurring carbohydrates exist as glycosidic compounds; they include oligo- or polysaccharides, glycolipids, glycoproteins, and nucleosides. In the biosynthesis of these compounds, the glycosidic bonds cannot be formed by the direct dehydration reactions of saccharide units and its aglycon parts. It is necessary to activate the anomeric center of the saccharide unit by introducing an appropriate leaving group so that the anomeric carbon atom is attacked by a hydroxyl group of the aglycon part (Shoda 2001). Catalysts responsible for the glycosylation of these activated saccharides are synthases, which are classified as glycosyl transferases. After being utilized, these glycosidic compounds are finally converted to carbon dioxide and water via combustion or degradation catalyzed by glycosides from bacteria. It is, therefore, obvious that two kinds of enzymes, glycosyl transferases and glycosidases, are involved in the process of glycosylation and deglycosylation in nature, constructing a large carbon cycle system.

Concept

A new carbon cyclic system which involves the production of functional glycosidic compounds such as artificial oligo- or polysaccharides based on the concept of "Glyco-chemistry cycle system" will be proposed (Fig. 1A) (Shoda et al. 2003). In this cyclic system, one glycosidase enzyme plays two roles including glycosidic bond formation and glycosidic bond cleavage. This concept consists of (1) the transformation of a polysaccharide biomass to refined raw materials, (2) the anomeric activation of the raw materials, (3) polymerization (glycosylation) catalyzed by a glycosidase to give functionalized oligosaccharides, (4) the degradation of the products by the glycosidase. This is a low environment-loading process because the characteristic feature of glycosidases is to catalyze both glycosylation reaction and deglycosylation reaction due to the complementarity of glycosidases and, therefore, the resulting products can be converted to the starting materials enzymatically without using any drastic reaction conditions or strong acid catalysts. The following sections deal with examples of glycosidase-catalyzed production of functionalized oligosaccharides.

Department of Biomolecular Engineering, Graduate School of Engineering, Tohoku University, 6-6-11-514, Aoba, Aoba-ku, Sendai 980-8579, Japan
Phone: +81-22-795-7230, Fax: +81-22-795-7293
E-mail: shoda@poly.che.tohoku.ac.jp

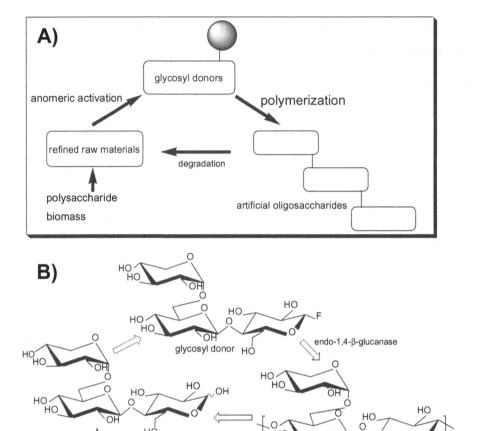

Fig. 1 Glcyco-chemistry cycle system based on glycosidases. **A** General concept. **B** Cyclic system based on xyloglucan as biomass source

Chemo-Enzymatic Synthesis of Xyloglucan Oligosaccharides

Naturally occurring xyloglucan in Tamarind seeds was trimmed off by the action of three kinds of enzymes, giving rise to xylopyranosylcellobiose, and the resulting trisaccharide was further transformed to the corresponding glycosyl fluoride monomer (Fig. 1B). When the fluoride monomer was treated with endo-1,4-β-glucanase (*Trichoderma reesei*) under a preferable condition for transglycosylation, i.e., in the presence of acetonitrile, a poly-condensation reaction took place, giving rise to artificial xyloglucan oligomers. It is impossible to construct such a structure using, the conventional method of modifying a naturally occurring polysaccharide by chemical reactions.

One-Pot Synthesis of Chitooligosaccharides by Combined Use of Mutant Chitinase and β-Galactosidase

A one-pot synthesis of (GlcNAc) ($n = 3$ and 4) has been demonstrated by combined use of a deactivated chitinase and β-galactosidase (Fig. 2) (Kohri et al. 2006). The oxazoline derivative (Gal-GlcNAc-oxa) was employed as glycosyl donor. First, the chitinase was

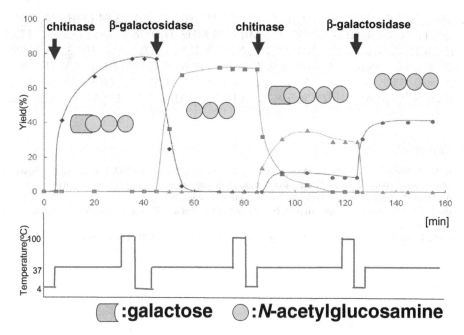

Fig. 2 One-pot synthesis of chitooligosaccharides by combined use of mutant chitinase and β-galactosidase

added to a mixture of Gal-GlcNAc-oxa and (GlcNAc)$_2$ at 4°C and the transglycosylation reaction was carried out at 37°C. The enzyme was inactivated and β-galactosidase was added at 4°C. Then, β-galactosidase-catalyzed degalactosylation of the transglycosylated product was performed. Finally, the β-galactosidase was inactivated at 100°C, giving rise to (GlcNAc)$_3$. By repeating this process, (GlcNAc)$_4$ was obtained without isolating any synthetic intermediates. All reactions were carried out in the same microtube, and the temperature was controlled using a thermal cycler.

Protocol: One-Pot Synthesis of Chitooligosaccharides

The process of GlcNAc unit elongation is as follows: (1) addition of a mutant chitinase (W433Y) to a mixture of Gal-GlcNAc-oxa and (GlcNAc)$_2$ at 4°C, (2) enzymatic trans-glycosylation reaction at 37°C for 30 min, (3) inactivation of the enzyme at 100°C for 15 min, (4) addition of β-galactosidase at 4°C, (5) β-galactosidase-catalyzed degalacto-sylation reaction at 37°C for 30 min, (6) inactivation of β-galactosidase at 100°C for 5 min [synthesis of (GlcNAc)$_3$], (7) addition of the mutant chitinase (W433Y) and Gal-GlcNAc-oxa to the mixture at 4°C, (8) enzymatic transglycosylation reaction at 37°C for 4 h (synthesis of Gal–GlcNAc–GlcNAc–GlcNAc–GlcNAc), (9) inactivation of the enzyme at 100°C for 15 min, (10) addition of β-galactosidase at 4°C, (11) β-galactosi-dase-catalyzed degalactosylation reaction at 37°C for 30 min, and (12) inactivation of β-galactosidase at 100°C for 15 min [synthesis of (GlcNAc)$_4$]. The reaction was termi-nated by adding acetonitrile to the reaction mixture and the yield of products was deter-mined by HPLC (eluant: water/acetonitrile = 1/3).

^{1}H NMR (500 MHz, D$_2$O) of pentasaccharide (Gal–GlcNAc–GlcNAc–GlcNAc–GlcNAc): δ 5.00 (d, $J_{1,2}$ = 2.4 Hz, H-1α), 4.51 (d, $J_{1,2}$ = 7.4 Hz, H-1β), 4.39–4.43 (3H,

m, H-1'/H-1″ /H-1‴), 4.29 (1H, d, $J_{1''',2'''}$ = 7.5 Hz, H-1‴″), 3.34–3.84 (30H, m), 1.88 ×
3, 1.86 (12H, s × 4, NHAc × 4); ^{13}C NMR (125 MHz, D_2O): δ 173.8, 173.7 × 2, 173.5
(NHCOCH_3^{1}β, NHCOCH_3^{2}, NHCOCH_3^{3}, NHCOCH_3^{4}), 102.0, 100.4, 100.3, 93.9, 89.5,
78.7, 78.2, 78.1, 78.0, 77.2, 74.4, 73.9, 73.7, 73.6, 71.6, 71.1, 70.0, 69.1, 68.3, 67.6, 60.1,
59.2, 59.0, 55.2, 54.2, 54.1, 52.7, 21.2 × 4 (NHCOCH_3^{1}β, NHCOCH_3^{2}, NHCOCH_3^{3},
NHCOCH_3^{4}), 21.0 (NHCOCH_3^{1}α); MALDI-TOF MS: 1,017.1 ([M + Na$^+$]; calc.
1,016).

References

Kohri M, Kobayashi A, Noguchi M, Kawaida S, Watanabe T, Shoda S (2006) Stepwise synthesis of
chitooligosaccharides through a transition-state analogue substrate catalyzed by nutants of chitinase
A1 from *Bacillus circulans* WL-12. Holzforschung 60:485–491
Shoda S (2001) Enzymatic glycosylation. In: Fraser-Reid B, Tatsuta K, Thiem J (eds) Glycoscience, vol
2. Springer, Heidelberg
Shoda S, Izumi R, Fujita M (2003) Green process in glycotechnology. Bull Soc Chem Jpn 76:1–13

Endoglycosidases (Glycosaminoglycans)

Ikuko Kakizaki, Keiichi Takagaki

Introduction

Currently established biotechnologies are mainly based on genetic engineering and protein engineering, and they provide a number of valuable recombinant proteins. These technologies have brought benefits to medicine by enabling the production of various drugs. However, genetically produced recombinant proteins often lack entirely or partially their native sugar chains or have sugar chains that are different from the original sugar chains, thus being apt to exert insufficient specific bioactivities or to be unstable. The cause of this lack of sugar chains of recombinant proteins is that, DNA contains genetic information only on amino acid sequences but not on the biosynthesis of sugar chains that link to proteins. This problem could be solved by technologies to attach sugar chains to sugar chain-deficient recombinant protein. Moreover, attaching artificially synthesized bioactive sugar chains to protein would make it possible to create glycoproteins with additional biological activity. We focused on the transglycosylation activity of endoglycosidases acting on proteoglycan and developed methods for the reconstruction of their sugar chains, glycosaminoglycan (GAG), and for attachment of GAGs to protein.

Concept

Glycosidases catalyze transglycosylation reaction as a reverse reaction in addition to their main hydrolysis reaction. In the case of transglycosylation using exoglycosidase, which hydrolyzes sugar chains by the stepwise removal of monosaccharide residues from the non-reducing termini, monosaccharide residues are transferred. In contrast, endoglycosidase, which hydrolyzes the internal glycosidic bonds of sugar chains, transfers oligosaccharides by transglycosylation reaction. Therefore, transglycosylation using endoglycosidase, i.e., "transfer reaction of sugar chains" is extremely valuable for artificial synthesis of sugar chains.

Proteoglycans are complex glycoconjugates composed of a core protein and one to more than 100 GAG chains, covalently bound to the core protein through a galactose–galactose–xylose–serin linkage region. Proteoglycan forms plasma membrane and extracellular matrix and shows various bioactivities in many biological aspects such as cell recognition, cell adhesion, and signal transduction by itself or by interacting with other molecules.

Two types of endoglycosidases; (1) enzymes acting on the glycosidic bonds at the inner region of the GAG chains and (2) enzymes acting on the linkage region between

Department of Biochemistry, Hirosaki University School of Medicine, 5 Zaifu-cho,
Hirosaki 036-8562, Japan
Phone: +81-172-39-5015, Fax: +81-172-39-5016
E-mail: kaki@cc.hirosaki-u.ac.jp

the core protein and the repeating disaccharide units of GAG, were discovered as proteoglycan-related endoglycosidases.

First, an enzymatic reconstruction method for GAGs was developed using one of the former type of endoglycosidases, testicular hyaluronidase, that acts on chondroitin sulfates in addition to hyaluronan. This enzyme transfers the disaccharide unit to the non-reducing terminal of an acceptor oligosaccharide. An example of the elongation of a hyaluronan chain at the non-reducing terminal from pyridylaminated hyaluronan hexasaccharide (PA-hyaluronan hexasaccharide) by transglycosylation of testicular hyaluronidase is described in "Protocol" section. We have constructed an oligosaccharide library of more than 100 octa- and decasaccharides to date using this enzymatic reconstruction method (Takagaki and Ishido 2000, Takagaki et al. 2000). Moreover, we have succeeded in applying these reconstructed oligosaccharides to obtain structural information of the functional domains on GAG and to determine a GAG sequence where certain GAG-degrading enzyme acts. We have also succeeded in the reconstruction of the GAG chain in decorin, which is the simplest proteoglycan, by the same method using transglycosylation of testicular hyaluronidase (Iwafune et al. 2002).

Endo-β-xylosidase, one of the latter enzymes, hydrolyzes xylosyl serine bonds of the linkage region resulting in the release of intact GAG chains from proteoglycans. Transglycosylation using this enzyme would enable the transfer of long chains of GAGs. It is possible that endoglycosidases could be used as a groundbreaking tool for glycotechnology, like restriction enzymes and ligases in gene manipulation. In order to use this tool for the synthesis of artificial proteoglycan, GAG chain carriers having the linkage region, such as 4-methylumbelliferyl-GAG, scheduled to be attached to a core protein, were prepared (Takagaki et al. 2002). We also have succeeded in introducing a GAG chain to a synthetic peptide using endo-β-xylosidase (Ishido et al. 2002). Figure 1 shows a future strategy of an enzymatic synthesis of artificial proteoglycan having a reconstructed GAG chain. These glycotechnologies using endoglycosidases must reinforce gene engineering and lead to the development of new biotechnology.

Protocol

Transglycosylation Reaction of Testicular Hyaluronidase

1. Incubate PA-hyaluronan hexasaccharide (≥2 nmol, acceptor) and hyaluronan (≥5 µg, average molecular weight 80,000, donor) with 1.0 NFU of bovine testicular hyaluronidase dissolved in 50 µl of 0.1 M Tris–HCl buffer of pH 7.0 at 37°C for 1 h.
2. Terminate the reaction by boiling for 5 min.

Analysis of the Transglycosylated Products by Normal Phase HPLC

1. Prepare solution A containing 3% acetic acid adjusted to pH 7.3 with triethylamine and acetonitrile at a ratio of 20:80 (v/v) and solution B containing the same agent at a ratio of 50:50 (v/v).
2. Equibrate a TSKgel Amide-80 column (4.6 × 250 mm, Tosoh Co., Tokyo, Japan) with solution A (set flow rate at 1.0 ml/min).
3. Elute sample on a linear gradient of 0–100% of solution B over 55 min. Follow the fluorescence of PA by detecting at excitation and emission wavelengths of 320 and 400 nm, respectively.

1. Preparation of oligosaccharide carrier as an acceptor

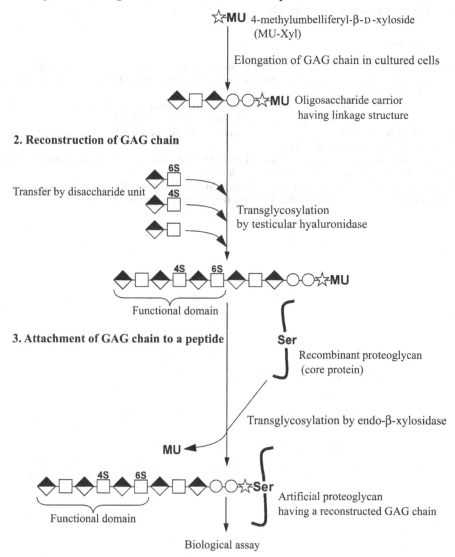

Fig. 1 Strategy of an enzymatic synthesis of artificial proteoglycan having a reconstructed glycosaminoglycan (GAG) chain

Comments

In the case of synthesis of an oligosaccharide having a hybrid structure consisting of more than three different disaccharide units, perform multi-step transglycosylation on the transglycosylated product of the previous step as an acceptor using a different kind of GAG as a donor for each step.

Change the reaction conditions of transglycosylation according to the donor/acceptor combinations.

Change the composition of the solutions A and B, and the gradient conditions according to the desired sugar chain length.

References

Ishido K, Takagaki K, Iwafune M, Yoshihara S, Sasaki M, Endo M (2002) Enzymatic attachment of glycosaminoglycan chain to peptide using the sugar chain transfer reaction with endo-beta-xylosidase. J Biol Chem 277:11889–11895

Iwafune M, Kakizaki I, Yukawa M, Kudo D, Ota S, Endo M, Takagaki K (2002) Reconstruction of glycosaminoglycan chains in decorin. Biochem Biophys Res Commun 297:1167–1170

Takagaki K, Ishido K (2000) Synthesis of chondroitin sulfate oligosaccharides using enzymatic reconstruction. Trends Glycosci Glyc 12:295–306

Takagaki K, Munakata H, Kakizaki I, Majima M, Endo M (2000) Enzymatic reconstruction of dermatan sulfate. Biochem Biophys Res Commun 270:588–593

Takagaki K, Ishido K, Kakizaki I, Iwafune M, Endo M (2002) Carriers for enzymatic attachment of glycosaminoglycan chains to peptide. Biochem Biophys Res Commun 293:220–224

Oligosaccharide Synthesis Based on Combinatorial Chemistry and Labo Automation

Hiroshi Tanaka, Takashi Takahashi

Introduction

Combinatorial chemistry is an effective method not only for the rapid assembly of small molecule libraries but also for the optimization of reaction conditions. Labo automation permits laborious and routine manipulations to be minimized. In this report, we describe the application of combinatorial chemistry and labo automation to the synthesis of oligosaccharide libraries.

One-Pot Glycosylation and Polymer-Assisted Deprotection

We developed two key methodologies for the synthesis of oligosaccharide libraries based on combinatorial chemistry and labo automation, namely one-pot glycosylation and polymer-assisted deprotection (Scheme 1). One-pot glycosylation involves the sequential activation of glycosyl donors **1–3** in a single vessel and circumvents the need for the isolation or purification of intermediate **4** to provide **5**. We investigated the one-pot glycosylation based on the chemo-selective activation of glycosyl donors having different leaving groups with appropriate activators (Yamada et al. 1994; Tanaka et al. 2002). The order of activation of glycosyl donors mainly relies on the nature of their leaving groups and activators. In addition, the requisite manipulations in one-pot glycosylation are adaptable to solution-phase automated synthesizers. Sialylation with *N*-Troc protected sialosides was also recently adapted for use in the reaction sequence for a one-pot glycosylation (Tanaka 2005b). On the other hand, polymer-assisted deprotection involves the deprotection of the solid-supported, protected oligosaccharides **6**. The solid-supported complex oligosaccharides **6** would smoothly undergo deprotection because they undergo aggregation to a very limited extent. A Birch reduction was adapted to remove the solid-supported benzyl ethers and esters. The THP linker can survive deprotection reactions and is cleaved under mildly acidic conditions, releasing the fully deprotected oligosaccharide **7** without anomerization or cleavage of the glycosidic bonds. The ease of handling of solid-supported compounds is effective not only for the high-speed synthesis of a single target oligosaccharide but also for the deprotection of a protected oligosaccharide library.

Automated Synthesis of Dimeric LewisX Derivatives by One-Pot Glycosylation

Scheme 2 shows a combinatorial synthesis of dimeric LewisX derivatives by the one-pot glycosylation utilizing an automated synthesizer (Tanaka et al. 2005a). The sequential

Department of Applied Chemistry, Graduate School of Science and Engineering, Tokyo Institute of Technology, 2-12-1 Ookayama, Meguro-ku, Tokyo 152-8552, Japan
Phone: +81-3-5734-2120, Fax: +81-3-5734-2884
E-mail: ttak@apc.titech.ac.jp,

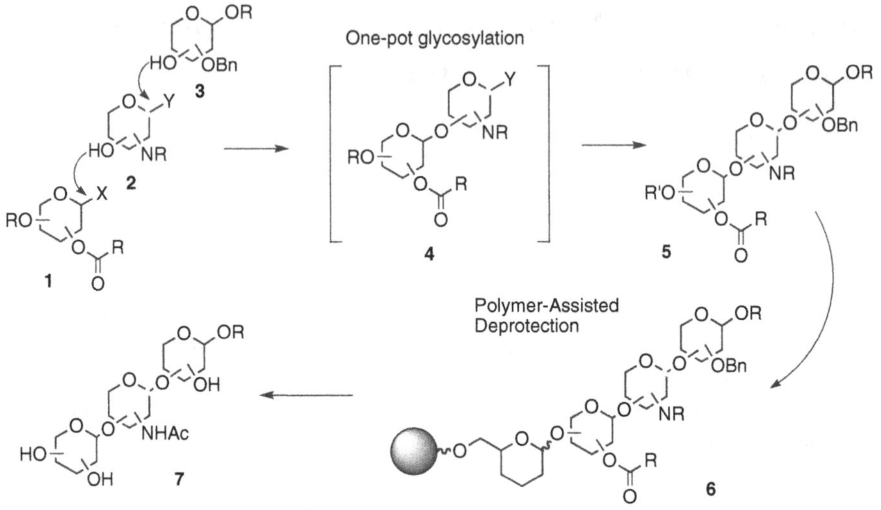

Scheme 1 One-pot glycosylation and polymer-assisted deprotection

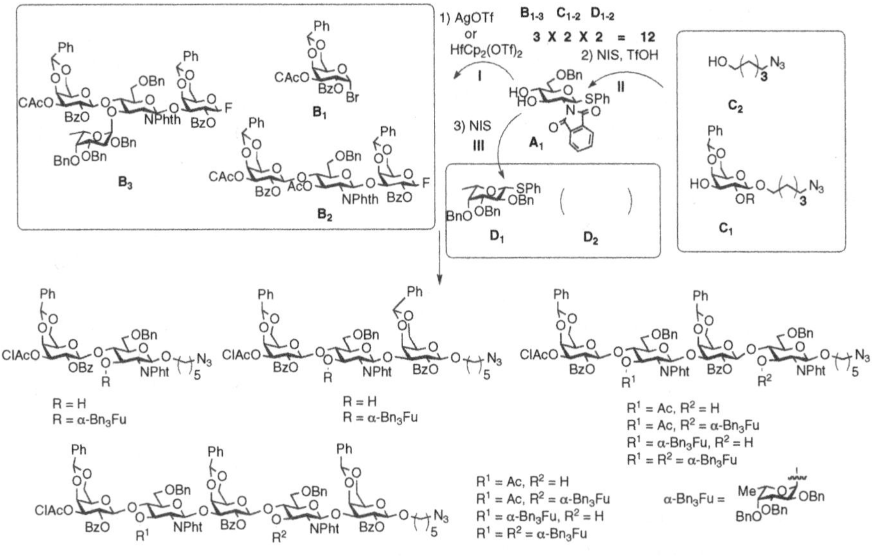

Scheme 2 A combinatorial synthesis of dimeric LewisX derivatives by one-pot glycosylation

three-step one-pot glycosylations using 12 sets of 4 building blocks A, B, C, and D (I → II → III) were achieved by utilizing an automated synthesizer (L-COS). Glycosyl bromide, B_1, and fluorides, B_2 and B_3, underwent chemo- and regio-selective glycosylation of thioglycoside A. The N-phthaloyl group of A reduces the reactivity of the C3 hydroxyl group of glucosamine A to promote regio-selective glycosylation at the C4 hydroxyl group, and assisted formation of the 1,2 trans glycosidic bond in the second glycosylation. The program schedule for the parallel synthesis of 12 oligosaccharides involves the order of addition of reagents, reaction times and reaction temperatures in

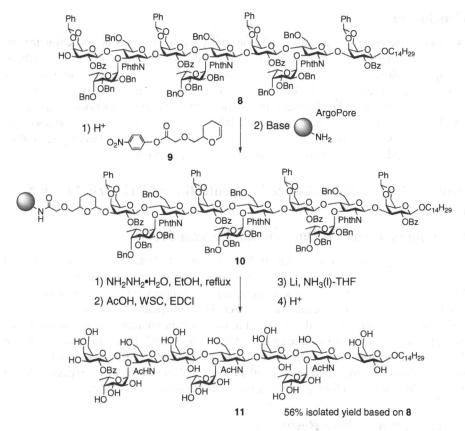

Scheme 3 Synthesis of the trimeric Lewis X derivative **11** based on the polymer-assisted deprotection method

each reaction vessel. Eight hours was required to complete the program after arranging all reagents. Purification of the 12 crude compounds by silica gel chromatography and gel permeation chromatography provided 12 protected oligosaccharides in 22–46% overall yields based on **C**.

Polymer-Assisted Deprotection of Protected Oligosaccharides

Scheme 2 shows an application of the polymer-assisted deprotection method to the synthesis of the trimeric Lewis X derivative **11** (Tanaka et al. 2006). The polymer-supported and -protected oligosaccharide **10** can be prepared using the following methodology: (1) acetal formation of the protected saccharides **8** with prelinker **9** containing a DHP moiety and an activated ester and (2) subsequent amidation of the resulting activated ester with the solid-supported amines to give **10**, irreversibly. The irreversible loading reaction enabled the complete immobilization of the protected oligosaccharides **8**. The solid-supported benzyl ethers were cleaved under Birch reduction conditions, in which an ArgoPore resin was critical for completing the deprotection reactions. Cleavage of the oligosaccharides under mildly acidic conditions provided the fully deprotected decasaccharide in a 56% yield based on **8**.

Conclusion

We have developed one-pot glycosylation and polymer-assisted deprotection for the synthesis of oligosaccharide libraries. Both methods ease the protocol for the synthesis of oligosaccharide and can be adapted to automated-synthesizers. In order to expand the applicability of the methods to the synthesis of various biologically active and complex oligosaccharides, the development of efficient glycosidation reactions such as α-sialylation, β-mannosylation, β- and α-selective glycosidation of 2-deoxyoligosaccharides continues to be needed. In addition, modification of oligosaccharide via sulfation and phospholylation is also an important issue that needs to be solved.

Protocol for the Solid-Assisted Deprotection of Trimeric Lewis X Derivative

Coupling of Protected Oligosaccharide with Prelinker

A mixture of protected trimeric Lewis X **8** (45.0 mg, 11.0 mmol, 1.0 eq.), prelinker **9** (9.8 mg, 33.0 μmol, 3.0 eq.) (azeotroped two times with toluene), and pulverized activated MS-4A in dry CH_2Cl_2 (0.22 mL) was stirred at room temperature for 15 min under argon atmosphere to remove a trace amount of water. Then, the reaction mixture was cooled to 0°C. 10-Camphorsulfonic acid was added to the reaction mixture. The reaction mixture was allowed to warm to room temperature. After being stirred at the same temperature for 2 h, the reaction mixture was filtered through a pad of Celite. The filtrate was concentrated in vacuo and chromatographed on a small amount of silica gel to remove excess prelinker **9**. The residue was used for the next reaction without further purification.

Loading on ArgoPore Resin

To a solution of the residue in dry CH_2Cl_2 (0.91 mL) and dry *N,N*-dimethylformamide (0.91 mL) was added ArgoPore resin (153 mg, 110 μmol, 10 eq.) and a catalytic amount of *N,N*-diisopropylethylamine at room temperature. After being shaken at room temperature for 12 h, the reaction mixture was filtrated and the resin was washed three times each with DMF (3.0 mL × 2), CH_2Cl_2 (3.0 mL × 2), MeOH (3.0 mL × 2), and CH_2Cl_2 (3.0 mL × 2). The resin was dried in vacuo to give the solid-supported, protected trimerc Lewis X epitope **10**. [FT-IR (solid) 3,024, 2,927, 1,719, 1,683, 1,602, 1,493, 1,451, 1,092, 1,028, 828, 759, 698 cm^{-1}.]

Modification of N-phtaloyl Group to N-acetyl Group

To a suspension of the solid-supported phthalimide **10** packed into MicroKans™ microreactor in ethanol (25.0 mL) was added hydrazine monohydrate (5.00 mL). After being stirred at 100°C for 6 h, the reaction mixture was filtered. The resin was washed three times each with ethanol (10 mL × 2), CH_2Cl_2 (10 mL × 2), THF/distillated water (1/1) (10 mL × 2), MeOH (10 mL × 2), and CH_2Cl_2 (10 mL × 2). The resin was dried in vacuo.

To a suspension of the resin in *N,N*-dimethylformamide (10.0 mL) was added acetic acid (62.0 μL, 1.10 mmol, 100 eq.), *N,N*-diisopropylcarbodiimide (170 μL, 1.10 mmol, 100 eq.), *N,N*-diisopropylethylamine (188 μL, 1.10 mmol, 100 eq.), and a catalytic amount of *N,N*-dimethyaminopyridine at room temperature. After being stirred at the same temperature for 6 h, the reaction mixture was filtered. The resin was washed three times each with CH_2Cl_2 (10 mL × 2), N,N-dimethylformamide (10 mL × 2), THF/distillated water (1/1) (10 mL × 2), MeOH (10 mL × 2), and CH_2Cl_2 (10 mL × 2). The resin was dried in vacuo to give the solid-supported acetamide [FT-IR (solid) 3,025, 2,920, 1,663, 1,602, 1,492, 1,451, 1,091, 827, 790, 704 cm^{-1}].

Debenzylation and Deacylation

The solid-supported acetamide was stirred in dry THF (5 mL) at room temperature for 15 min. Then, liq. NH_3 (45 mL) and lithium-granule (50 mg) were added to the reaction mixture at −78°C. After being stirred at the same temperature for 1.5 h, the reaction mixture was allowed to warm to reflux. After being stirred under reflux for 1 h, the reaction mixture was quenched with MeOH and stirred at room temperature for 12 h. The reaction mixture was filtered and the resin was washed three times each with THF (10 mL × 2), CH_2Cl_2 (10 mL × 2), THF/distillated water (1/1) (10 mL × 2), MeOH (10 mL × 2), and CH_2Cl_2 (10 mL × 2). The resin was dried in vacuo.

Release from Resin

The resin was treated with a solution of trifluoroacetic acid (0.80 mL), MeOH (0.10 mL), and CH_2Cl_2 (3.1 mL) at room temperature for 15 min. The reaction mixture was filtered and rinsed with CH_2Cl_2 and MeOH. The filtrate was concentrated and azeotroped three times with toluene. The cleavage purity was >99% and analyzed by reversed phase HPLC [Waters® ODS-UG-5 column, eluent: 0.1% TFA in H_2O–0.1% TFA in THF, gradient: 0.00 min (0.1% TFA in H_2O):(0.1% TFA in THF) = 65:35, 8.00 min (0.1% TFA in H_2O):(0.1% TFA in THF) = 65:35, 1 mL min^{-1}; retention time: 5.77 min]. The residue was purified by reverse-phase column chromatography (Bond Elut-C18) to give fully deprotected trimeric Lewis X **11** (12.1 mg, 6.3 mmol, 58% based on **8**). [a]$_D$^{24} −47.0° (c 0.550, MeOH); ^{1}H NMR (400 MHz, CD$_3$OD) δ 5.05 (m, 3H), 4.80 (m, 3H), 4.68 (m, 3H), 4.43 (m, 3H), 4.19 (d, 1H, J = 7.7 Hz), 4.04 (m, 1H), 3.99–3.39 (m, 52H), 1.96 (m, 9H), 1.60 (m, 2H), 1.40–1.22 (m, 22H), 1.15, (m, 9H), 0.89 (t, 3H, J = 6.8 Hz); ^{13}C NMR (100 MHz, CD$_3$OD) δ 175.4, 175.4, 175.2, (105.7, 105.7, 104.6, 104.5, 104.5, 104.5, 104.4, 101.0, 101.0, 100.9, anomeric), 84.8, 84.7, 78.0, 77.4, 77.2, 77.1, 76.9, 76.8, 75.7, 75.6, 74.4, 74.4, 73.5, 72.7, 72.3, 71.9, 71.6, 70.7, 70.7, 70.6, 70.6, 70.2, 68.4, 63.6, 63.5, 63.1, 62.0, 58.4, 33.8, 31.6, 31.5, 31.5, 31.4, 31.2, 27.9, 24.5, 24.0, 17.4, 15.2; FT-IR (neat) 3,317, 2,926, 1,646, 1,566, 1,385, 1,071, 891, 718, 529 cm^{-1}; HRMS (ESI-TOF) Calcd for $C_{80}H_{139}N_3O_{48}Na$ [M + Na]$^+$ 1932.8420, found 1932.8425.

References

Tanaka H, Adachi M, Tsukamoto H, Ikeda T, Yamada H, Takahashi T (2002) Synthesis of Di-branched heptasaccharide by one-pot glycosylation using seven independent building blocks. Org Lett 4:4213–4216

Tanaka H, Matoba N, Tsukamoto H, Takimoto H, Yamada H, Takahashi T (2005a) Automated parallel synthesis of a protected oligosaccharide library based upon the structure of dimeric Lewis X by one-pot sequential glycosylation. Synlett 824–828

Tanaka H, Adachi M, Takahashi T (2005b) One-pot synthesis of sialo-containing glycosyl amino acids using N-Troc β-thiophenyl sialoside. Chem Eur J 11:849–862

Tanaka H, Ishida T, Matoba N, Tsukamoto H, Yamada H, Takahashi T (2006) Polymer-assisted strategy for the deprotection of protected oligosaccharides. Angew Chem Int Ed 45:6349–6352

Yamada H, Harada H, Takahashi T (1994) Synthesis of an elicitor-active hexaglucoside analog by a one-pot, 2-step glycosidation procedure. J Am Chem Soc 116:7919–7920

Endoglycosidases (Glycoproteins)

Kaoru Takegawa,[1] **Kenji Yamamoto**[2]

Introduction

Endo-β-N-acetylglucosaminidase (EC 3.2.1.96) hydrolyzes the glycosidic bond in the *N, N'*-diacetylchitobiose moiety of *N*-linked oligosaccharides of various glycoproteins. This enzyme is unique with regard to the enzymatic action by which one GlcNAc residue remains bound to the protein/peptide.

R-GlcNAc-GlcNAc-Asn-Protein (or peptide) + H₂O

↓ ←endo-β-*N*-acetylglucosaminidase

R-GlcNAc + GlcNAc-Asn-Protein (or peptide) (R: oligosaccharide)

Endo-β-*N*-acetylglucosaminidase is one of the most well-studied endoglycosidases used for the structural analysis of glycoproteins and has been used for isolation of both *N*-linked oligosaccharides and partially deglycosylated proteins without damaging the glycoproteins.

Novel endo-β-*N*-acetylglucosaminidases from *Arthrobacter protophormiae* (Endo-A) and *Mucor hiemalis* (Endo-M) show transglycosylation activity and can transfer the oligosaccharides from glycoprotein (glycoside donor) to suitable acceptors with a GlcNAc residue during hydrolysis of the *N*-linked oligosaccharides (Yamamoto and Takegawa 1997). Endo-A specifically acts on high mannose-type oligosaccharides, whereas Endo-M acts on not only high mannose-type oligosaccharides but also on hybrid-type and complex-type oligosaccharides. Endo-β-*N*-acetylglucosaminidases are classified into two glycoside hydrolase (GH) families of 18 and 85 from their structures. Endo-A (Takegawa et al. 1997) and Endo-M (Fujita et al. 2004) belong to GH family 85, and their amino acid sequences are different from those of GH family 18 enzymes such as Endo-H from *Streptomyces plicatus*, which does not have transglycosylation activity. Interestingly, GH family 85 enzymes can be found in multiple eukaryotic organisms, such as worm, plant, fly, and human. The transglycosylation activity of endo-β-*N*-acetylglucosaminidases might positively participate in various cellular functions in eukaryotic cells.

Although peptides can be easily synthesized by a chemical method, it is very difficult to synthesize glycopeptides chemically not only because the chemical synthesis of oligosaccharide involves complicated steps, but also because of the difficulty in binding it

[1] Faculty of Agriculture, Kagawa University, 2393 Ikenobe, Miki-cho, Kagawa 761-0795, Japan
Phone: +81-87-891-3116, Fax: +81-87-891-3021
E-mail: takegawa@ag.kagawa-u.ac.jp

[2] Graduate School of Biostudies, Kyoto University, Kitashirakawa, Sakyo-ku, Kyoto 606-8502, Japan
Phone: +81-75-753-6277, Fax: +81-75-753-9228
E-mail: yamamotk@kais.kyoto-u.ac.jp

with the peptide. A novel method for the enzymatic synthesis of neoglycoproteins using oligosaccharide-transferring activity of endo-β-N-acetylglucosaminidase
has been developed. The method is suitable and practical for the improved synthesis of
neoglycopeptides and neoglycoproteins and useful for attaching the same N-linked
oligosaccharide chains to all the original N-glycosylation sites of glycoprotein
molecules.

Principle and Applications

Figure 1(A) illustrates the hydrolytic and transglycosylation activities of endo-β-N-
acetylglucosaminidase. After forming the enzyme–substrate complex, the enzyme could
transfer the donor oligosaccharides to either water (leading to hydrolysis) or an acceptor
(leading to transglycosylation). Since the transglycosylation product can be the substrate
for the same enzyme, the process will be repeated until all substrates are transferred to
water (complete hydrolysis). Therefore, the transglycosylation product is predominantly
formed in the initial stage of the reaction. As the reaction proceeds, the same enzyme
rapidly hydrolyzes the product; the reaction conditions should be carefully attained to
obtain optimal transglycosylation activity.

A method for the conversion of heterogeneous to homogeneous N-linked
oligosaccharide chains in glycoproteins, utilizing the transglycosylation activity of
Endo-A and Endo-M, is illustrated in Fig. 1(B). Partially deglycosylated glycoprotein
(GlcNAc-glycoprotein) is isolated by endo-β-N-acetylglucosaminidase digestion, and
the GlcNAc-containing glycoprotein is used as the acceptor of the transglycosylation
reaction.

In general, naturally occurring glycoproteins have a high degree of heterogeneity in
their oligosaccharide moiety, and it is difficult to clarify the biological roles of individual
oligosaccharides. Bovine pancreas ribonuclease B has one sugar chain in a heterogeneous
form. $(Man)_6(GlcNAc)_2Asn$ is incubated with Endo-A in the presence of partially deglycosylated ribonuclease B (GlcNAc-ribonuclease B). The molecular weight of the ribonuclease B is increased, and it migrates to the same position as the native enzyme,
indicating that the $(Man)_6GlcNAc$ has been transferred to the GlcNAc-ribonuclease. The
heterogeneous N-linked oligosaccharide chains, $(Man)_{5-9}(GlcNAc)_2$, of the native ribonuclease B are converted into homogeneous $(Man)_6(GlcNAc)_2$ by Endo-A (Takegawa
and Fan 2003). The high mannose-type oligosaccharide of ribonuclease B has also been
converted to the complex-type oligosaccharides by Endo-M.

Bioactive peptides containing N-linked oligosaccharides are chemo-enzymatically
synthesized by use of the solid-phase method of peptide synthesis and the transglycosylation activity of endo-β-N-acetylglucosaminidase (Haneda et al. 2003). Eel calcitonin is
a calcium-regulating peptide hormone composed of 32 amino acid residues and is
not naturally glycosylated. N-Glycopeptides containing a GlcNAc moiety have been
made by a dimethylphosphinothioic mixed anhydride method without protection of the
hydroxyl function of the sugar moiety. Complex-type oligosaccharides are introduced
to the calcitonin derivatives that contained GlcNAc residues at different sites (Asn3,
Gln14, and Gln20) by treatment with Endo-M. Using this chemo-enzymatic method,
oligosaccharides to the Asn and Gln residues of non-glycosylated peptides could be
introduced. Here, we present a typical method for remodeling of the sugar chains in
glycoproteins.

(A)

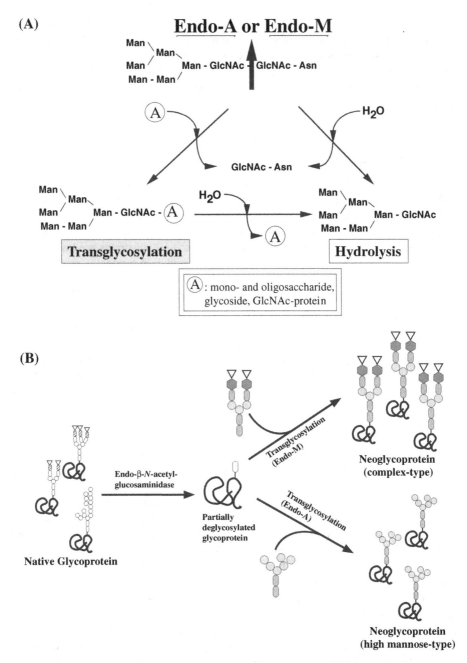

Fig. 1 A A putative mechanis of transglycosylation activity of endo-β-*N*-acetylglucosaminidase.
B Enzymatic synthesis of neoglycoproteins by transglycosylation with endo-β-*N*-acetylglucosaminidase.
○, mannose; ▯, *N*-acetylglucosaminine; ⬡, galactose; ▽, *N*-acetylneuraminic acid

Protocol (Remodeling of Sugar Chains by Endo-M)

1. Bovine pancreas ribonuclease B (RNase B, Sigma Chemicals Co.) has one sugar chain of the high mannose-type. First, 50 μg of RNase B and 0.1 mU of Endo-H in a total volume of 500 mL with 50 mM phosphate buffer (pH 5.5) is incubated overnight at 37°C to obtain GlcNAc-RNase B.

2. The reaction mixture is loaded onto a gel filtration column (Superdex 75 HR), and the fractions including GlcNAc-RNase B are concentrated and desalted on by Ultrafree-15 centrifugal filters 10K (Millipore).

3. Recombinant Endo-M (0.5 mU)(A1651, Tokyo Kasei Co. Ltd., Japan) is incubated with 3.5 mM sialyl glycopeptide from hen egg yolk, which contains sialobianntenary complex type oligosaccharide, in 40 μL of 25 mM acetate buffer (pH 6.0) in the presence of 40 μg of GlcNAc-RNase B.

4. After incubation of the mixture for 60 min at 30°C, the reaction is stopped by boiling for 3 min, and the reaction mixtures are analyzed by SDS-PAGE. The neo-RNase is detected by SSA (*Sambucus sieboldiana* agglutinin which is specific for sialic acid) blotting. Thus, the high mannose-type oligosaccharides of RNase B has been converted to the complex-type oligosaccharides.

References

Fujita K, Kobayashi K, Iwamatsu A, Takeuchi M, Natsuka S, Ikura K, Kumagai H, Yamamoto K (2004) Molecular cloning and characterization of *Mucor hiemalis* endo-β-*N*- acetylglucosaminidase that can transglycosylate complex type N-linked oligosaccharides. Arch Biochem Biophys 432:41–49

Haneda K, Inazu T, Mizno M, Yamamoto K (2003) Chemoenzymatic synthesis of neoglycoconjugates by transglycosylation with endo-β-*N*-acetylglucosaminidase from *Mucor hiemalis*. Methods Enzymol 362:74–85

Takegawa K, Fan JQ (2003) Enzymatic synthesis of neoglycoconjugates by transglycosylation with endo-β-*N*-acetylglucosaminidase A. Methods Enzymol 362:64–74

Takegawa K, Yamabe K, Fujita K, Tabuchi M, Mita M, Watanabe A, Asada Y, Sano M, Kondo A, Kato I, Iwahara S (1997) Cloning, sequencing, and expression of *Arthrobacter protophormiae* endo-β-*N*-acetylglucosaminidase in *Escherichia coli*. Arch Biochem Biophys 338:22–28

Yamamoto K, Takegawa K (1997) Transglycosylation activity of microbial endoglycosidases and its application. Trends Glycosci Glycotechnol 9:339–354

Glycosaminoglycans

Jun-ichi Tamura

Introduction

Glycosaminoglycans (GAGs), the glycan part of the proteoglycan, are classified into several categories based on the type of the so-called "repeating disaccharide" moieties which consist of hexosamine and hexuronic acid. The repeating disaccharide region occupies most of the GAG. A variety of the GAG biological activities is controlled by the complex, but a minute structure of the repeating oligosaccharides including the regiospecific sulfate groups. GAGs obtained from nature have heterogeneous structures. On the other hand, chemically synthesized GAGs have well-defined structures in a molecular level, and sufficient amounts can be supplied. To study the chemical, biological, medical as well as physical performance of the GAGs, a synthetic approach is strongly recommended. This chapter describes the synthesis of the GAG oligosaccharides focusing on the repeating region.

Synthetic Procedures

Many types of GAG repeating oligosaccharides have been synthesized by several groups, of which the details have been described in a few reviews (Tamura 2001; Karst and Linhardt 2003). Normally, GAG repeating oligosaccharides are synthesized in a systematic way using common disaccharide units. In most cases, the "disaccharide units" are indispensable for the effective oligosaccharide syntheses. The disaccharide units contain suitable protecting groups for further manipulations. Conventional coupling techniques using a halosugar, thioglycoside, as well as trichloroacetimidate, etc., can be adopted. The synthetic strategy should be well examined. Especially, the selections of the protecting groups for the regio- and stereoselective glycosylations, oxidation at the C-6 of Glc to Ido, and the sulfation at specific positions are important. Particularly, the synthesis of GAGs containing the α-L-IdoA moiety, i.e., dermatan sulfate, heparin, and heparan sulfate, requires complex strategies.

Heparin and heparan sulfate (HS) are composed of GlcNAc-GlcA/IdoA-type disaccharide moieties. GlcNSO$_3$Na often replaces GlcNAc. Many variations in the sulfation pattern of the OH and NH$_2$ complicate the structure of heparin and HS. The syntheses of heparin and HS have been started by Sinaÿ et al. The same group synthesized heparin pentasaccharide and the isomers which interacted with antithrombin III. van Boeckel's group has been synthesizing a number of HS analogues for practical use.

The hexasaccharide of dermatan sulfate, [GalNAc(4S)-IdoA(2S)]$_3$, was precisely synthesized by Ogawa and his co-workers. It is noteworthy that the hexasaccharide was proved to have a remarkable interaction with the heparin cofactor II.

Department of Regional Environment, Tottori University, Tottori 680-8551, Japan
Phone: +81-857-31-5108, Fax:+81-857-31-5108
E-mail: jtamura@rstu.jp

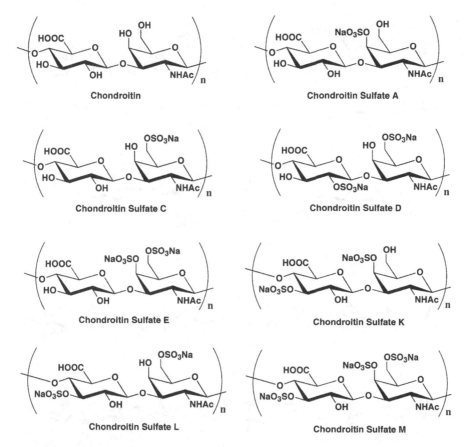

Fig. 1 Structural diversity of chondroitin sulfate disaccharide units

Hyaluronic acid (HA) is composed of a simple repeating disaccharide unit (GlcNAc-GlcA) without sulfate. Several groups have studied the synthesis of HA using different strategies. Jacquinet's group synthesized the octasaccharide (GlcNAc-GlcA)$_4$ as a methyl glycoside in a systematic way.

Recently, chondroitin sulfates (CSs) are becoming interesting topics in the medical and biological fields. CS contains the GalNAc-GlcA type repeating disaccharide being classified into several types based on the sulfation patterns as shown in Fig. 1. Ogawa's group has synthesized the A, C, and E types of CS for each di-, tri-, and tetrasaccharide. We have synthesized the hexa- and octasaccharides of CS-E by a stepwise glycan elongation. We adopted an acetamide-type GalNAc-GlcA building block and employed an effective strategy for which the coupling yields were high even for the longer oligosaccharide synthesis (Scheme 1) (Tamura and Tokuyoshi 2004; Tamura et al. 2008, see the protocol below). Jacquinet and his co-workers have been studying the synthesis of the CSs containing the GlcA–GalNAc type sequence. They succeeded in increasing the CS-D up to hexasaccharide. The naturally occurring CS can be utilized as a source of the disaccharide unit having the same sequence. Jacquinet's group demonstrated the acid hydrolysis of the CS polymer and obtained the disaccharide, which was converted into the corresponding chondroitin hexasaccharide (Lopin and Jacquinet 2006). Recently,

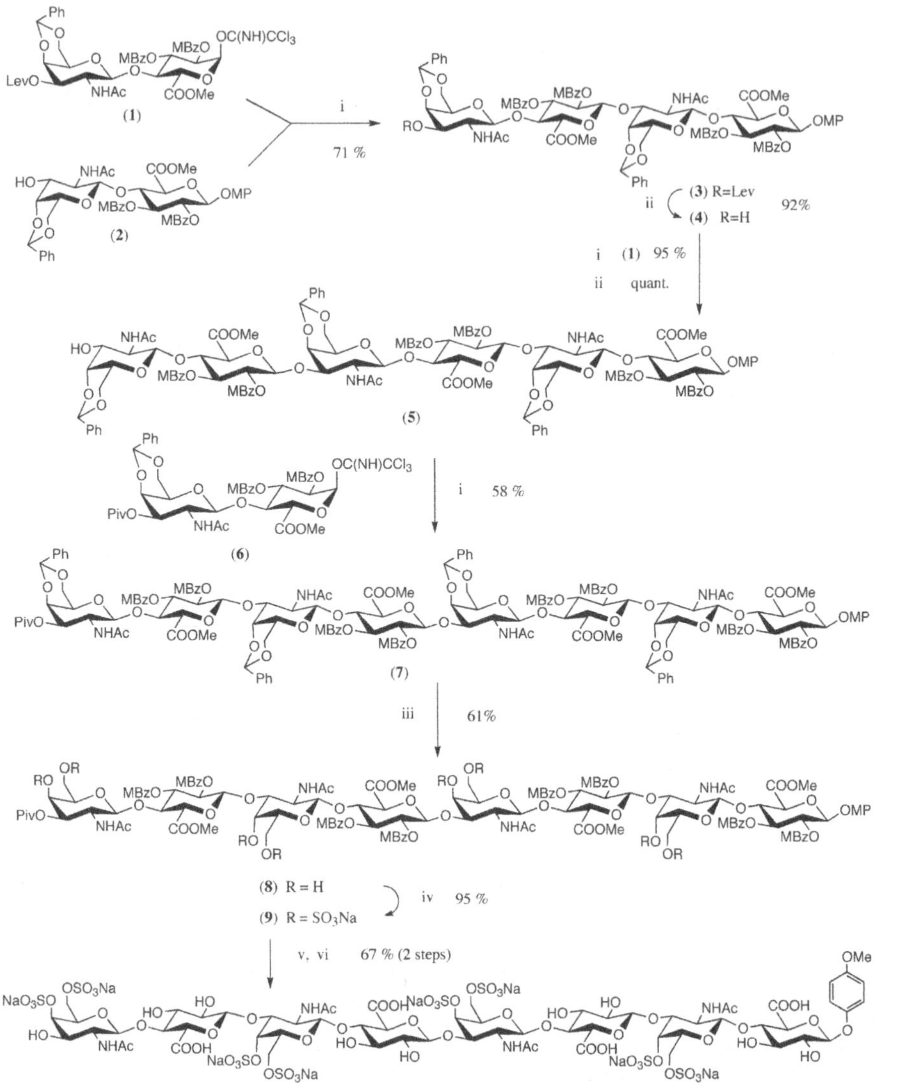

Chondroitin sulfate E octasaccharide

Abbreviations: Lev, -C(=O)(CH_2)$_2$COMe; MBz, -C(=O)C_6H_4Me(p); MP, -C_6H_4OMe(p); Piv, -C(=O)CMe_3

Reaction conditions: i, TMSOTf, MSAW300 / CH_2Cl_2, -20°C~rt,; ii,H_2NNH_2•AcOH / EtOH-toluene; iii, camphorsulfonic acid / CH_2Cl_2-MeOH; iv, SO_3•Me_3N / DMF, 60 °C; v, 1.25 M LiOH/ THF-H_2O; vi, 0.5 M NaOH

Scheme 1 Synthesis of chondroitin sulfate octasaccharide

Hsieh-Wilson's group synthesized the CS-E tetrasaccharide having the GlcA–GalNAc type sequence which stimulated the outgrowth of hippocampal neurons (Tully et al. 2004). They also established an analytical technique to detect the interaction between a specific protein, such as TNF-α, and the synthesized CS oligosaccharides of which the reducing terminal is bound to the surface of microarrays (Tully et al. 2006).

Further applications with the synthesized GAG oligosaccharides will be available in the fields of chemistry, biology, biochemistry as well as medicine.

Protocol: Synthesis of Chondroitin Sulfate E Octasaccharide

Glycan Elongation

To a solution of **6** (233.2 mg, 0.242 mmol) and **5** (277.2 mg, 0.122 mmol) in CH_2Cl_2 (10 mL) was added MSAW300 (580 mg). The mixture was stirred for 1 h at room temperature and then cooled to $-20°C$. TMSOTf (44 µL, 0.24 mmol) was added while continuing stirring for 1 day and gradually increasing to room temperature. The reaction was worked up by the addition of excess amount of Et_3N (67 µL) and aqueous $NaHCO_3$. $CHCl_3$ was added to the mixture, and the insoluble materials were removed by filtering through Celite. The organic phase was washed with aqueous $NaHCO_3$ and brine, and dried over anhydrous $MgSO_4$. The crude materials obtained were passed through gel permeation (LH-20, $CHCl_3$-MeOH 1:1) to give **7** (219.0 mg) in 58% yield. Rf 0.73 (EtOAc-MeOH 40:1). $[\alpha]_D$ +11.6° (c 0.88, $CHCl_3$). NMR δ_H (500 MHz, $CDCl_3$): 7.97 (2H, d, $J = 8.24$ Hz, Ar), 7.92 (4H, m, Ar), 7.87–7.79 (12H, m, Ar), 7.66 (2H, d, $J = 7.56$ Hz, Ar), 7.56 (2H, m, Ar), 7.45 (2H, m, Ar), 7.40–7.35 (3H, m, Ar), 7.32–7.24 (11H, m, Ar), 7.21–7.12 (12H, m, Ar), 7.06 (2H, d, $J = 8.25$ Hz, Ar), 6.89–6.87 (2H, m, Ar), 6.76–6.73 (2H, m, Ar), 5.72 (1H, t, $J_{2,3} = J_{3,4} = 8.94$ Hz, H-3^1), 5.70 (1H, s, PhCH), 5.58 (4H, m, H-3^3, 3^5, 3^7, PhCH), 5.47 (1H, dd, $J_{1,2} = 7.10$ Hz, H-2^1), 5.45 (1H, s, PhCH), 5.42 (1H, d, $J_{2,NH} = 6.88$ Hz, NH^8), 5.29 (1H, s, PhCH), 5.17 (3H, m, H-1^1, 2$^{3/5/7}$, 1^8), 5.04 (1H, brt, $J = 4.01$ Hz, H-2$^{3/5/7}$), 4.96 (4H, m, H-1^3, 1^5, 1^7, 2$^{3/5/7}$), 4.86 (1H, dd, $J_{3,4} = 7.33$, $J_{4,5} = 10.77$ Hz, H-4^7), 4.73 (2H, m, H-4$^{3/5}$, NH$^{2/4/6}$), 4.67–4.58 (3H, m, H-1$^{2/4/6}$, 4$^{5/3}$, 3^8), 4.50 (1H, t, H-4^1), 4.49 (1H, d, $J_{3,4} = 4.12$ Hz, H-4$^{2/4/6}$), 4.40 (1H, d, $J_{3,4} = 3.21$ Hz, H-4$^{2/4/6}$), 4.38 (1H, d, $J_{2,NH} = 8.70$ Hz, NH$^{2/4/6}$), 4.32 (1H, d, H-5^7), 4.30 (1H, d, $J_{3,4} = 3.44$ Hz, H-4^8), 4.27 (1H, d, $J_{4,5} = 10.31$ Hz, H-5$^{3/5}$), 4.19 (3H, brd, $J = 9.07$ Hz, H-5^1, 5$^{5/3}$, 1$^{2/4/6}$), 4.12 (3H, brd, $J = 4.36$ Hz, H-1$^{2/4/6}$, 4$^{2/4/6}$, NH$^{2/4/6}$), 3.99–3.78 (7H, m, H-2$^{2/4/6}$x2, 3$^{2/4/6}$x2, 6$^{2/4/6}$x2, 6b^8), 3.75–3.49 (7H, m, H-2$^{2/4/6}$, 3$^{2/4/6}$, 6$^{2/4/6}$x4, 6a^8), 3.73, 3.70, 3.68, 3.64, 3.64 (3Hx5, 5s, 4COOMe, MeOPh), 3.22 (1H, m, H-2^8), 2.97 (1H, s, H-5^8), 2.78. 2.68, 2.49 (1Hx3, 3s, H-5^2, 5^4, 5^6), 2.44, 2.41, 2.40, 2.39, 2.36, 2.34, 2.30 (24H, 7s, 8MePh), 1.75, 1.68, 1.67, 1.63 (3Hx4, 4s, 4NAc), 1.07 (9H, s, tert-Bu). Anal. calcd. for $C_{164}H_{172}N_4O_{55} \cdot 3H_2O$: C, 62.86; H, 5.74; N, 1.79%. Found: C, 62.42; H, 5.48; N, 1.44%.

Sulfation and Deprotection

Camphorsulfonic acid (28.2 mg, 121 µmol) was added to a solution of **7** (55.6 mg, 18.0 µmol) in CH_2Cl_2 (4.2 mL) and MeOH (4.2 mL) while stirring. After stirring for 22 h excess amount of Et_3N was added to the reaction mixture, and the volatiles were removed under reduced pressure. The residue was subjected to gel permeation chromatography (LH-20 $CHCl_3$-MeOH 1:1) to give **8** (30.3 mg) in 61% yield. ^{1}H-NMR measurement indicated complete disappearance of the four benzylidene acetals. To a solution of **8** (30.3 mg, 11.1 µmol) in DMF (2 mL) was added $SO_3 \cdot Me_3N$ (246 mg, 1.77 mmol) while stirring at 60°C for 3 days. The reaction mixture was then cooled to room temperature and subjected directly to gel permeation chromatography (LH-20 $CHCl_3$-MeOH 1:1) and an ion-exchange resin column [Dowex AG50W-X8 (Na^+), MeOH-H_2O 8:1] to give **9** (37.4 mg) in 95% yield. This compound was used for the next reaction without further purification. To a solution of **9** (37.4 mg, 10.5 µmol) in THF (2.2 mL) and H_2O (0.3 mL) was added 1.25 M LiOH (0.6 mL) at 0°C. After stirring overnight, all the volatiles were removed *in vacuo*. The residue was dissolved in MeOH (1.6 mL) and CH_2Cl_2 (0.3 mL), to which was added 0.5 M NaOH (1 mL) dropwise while stirring at room temperature. After 18 h, the reaction was quenched with 50% acetic acid. All the volatiles were removed under reduced pressure, and the residue was subjected to gel permeation chromatography (LH-20 1% AcOH) to afford the target **Chondroitinsulfate E Octasaccharide** (18.0 mg) in 70% yield. Rf 0.32 (RP-TLC, CH_3CN-H_2O 82:18). $[\alpha]_D$ +4.9 (*c* 0.61, H_2O). NMR δ_H (500 MHz, D_2O, internal tert-BuOH): 7.13–7.09 (2H, m, Ar), 6.99–6.97 (2H, m, Ar), 5.09 (1H, d, $J_{1,2} = 8.02$ Hz, H-1^1), 4.79 [3H, 3s, H-4 (GalNAcx3)], 4.72 [1H, s, H-4 (GalNAc)], 4.68 [1H, d, $J_{1,2} = 7.10$ Hz, H-1 (GalNAc)], 4.66 [1H, d, $J_{1,2} = 6.87$ Hz, H-1(GalNAc)], 4.62–6.55 [5H, m, H-1 (GlcAx3, GalNAcx2)], 4.30–4.19 [8H, m, H-6ab (GalNAcx4)], 4.16 (1H, d, $J_{4,5} = 9.62$ Hz, H-5^1), 4.11–4.07 [4H, m, H-5 (GalNAcx4)], 4.09–4.06 [3H, m, H-3 (GalNAcx3)], 4.03–3.97 [6H, m, H-5 (GlcAx3), H-2 (GalNAcx3)], 3.94 (1H, t, H-4^1), 3.89 [2H, m, H-2, 3 (GalNAc)], 3.89–3.83 [3H, m, H-4 (GlcAx3)], 3.81 (3H, s, *Me*OPh), 3.79 (1H, m, H-3^1), 3.68–3.64 [3H, m, H-3 (GlcAx3)], 3.64 (1H, m, H-2^1), 3.44–3.40 [3H, m, H-2 (GlcAx3)], 2.02, 2.00, 1.99, 1.99 (4H, 4s, 4NAc). MS (ESI): *m/z* (M-3H)$^{3-}$ calcd. for $C_{63}H_{81}N_4O_{70}S_8Na_8$, 817.66; found, 817.65; (M-Na-2H)$^{3-}$ calcd. for $C_{63}H_{82}N_4O_{70}S_8Na_7$, 810.33; found, 810.32; (M-2Na-H)$^{3-}$ calcd. for $C_{63}H_{83}N_4O_{70}S_8Na_6$, 803.01; found, 803.00; (M-3Na)$^{3-}$ calcd. for $C_{63}H_{84}N_4O_{70}S_8Na_5$, 795.68; found, 795.67.

References

Karst NA, Linhardt RJ (2003) Recent chemical and enzymatic approaches to the synthesis of glycosaminoglycan oligosaccharides. Curr Med Chem 10:1993–2031

Lopin C, Jacquinet J-C (2006) From polymer to size-defined oligomers: an expeditious route for the preparation of chondroitin oligosaccharides. Angew Chem Int Ed 45:2574–2578

Tamura J (2001) Recent advances in the synthetic studies of glycosaminoglycans. Trend Glycosci Glycotech 13: 65–88

Tamura J, Tokuyoshi M (2004) Synthesis of chondroitin sulfate E hexasaccharide in the repeating region by an effective elongation strategy toward longer chondroitin oligosaccharide. Biosci Biotech Biochem 68:2436–2443

Tamura J, Nakada Y, Taniguchi K, Yamane M (2008) Synthesis of chondroitin sulfate E octasaccharide in a repeating region involving an acetamide auxiliary. Carbohydr Res 343:39–47

Tully SE, Mabon R, Gama CI, Tsai SM, Liu X, Hsieh-Wilson LC (2004) A chondroitin sulfate small molecule that stimulates neuronal growth. J Am Chem Soc 126:7736–7737

Tully SE, Rawat M, Hsieh-Wilson LC (2006) Discovery of a TNF-α antagonist using chondroitin sulfate microarrays. J Am Chem Soc 128:7740–7741

Recent Advances in the Production of Mammalian-Type Sugar Chains in Yeast

Yasunori Chiba,[1,2] Yoshifumi Jigami[1]

Introduction

Protein therapeutics, such as enzyme replacement therapy, antibody therapeutics, and cytokine administration, are now known as the largest class of new candidates developed by the pharmaceutical companies. Yeasts have been used to produce industrial enzymes and are often chosen as the expression system because manufacturing costs are of primary concern. However, yeasts have a drawback of inability to attach mammalian-type sugar chain for the production of therapeutic glycoproteins for human use. N-glycosylated sugar chain in yeast is a mannan-type, which is partly antigenic in human and is sometimes trapped and cleared by mannose-specific receptors or lectins. Over the past 15 years or so, several approaches have been attempted to substitute yeast glycosylation pathway for a human one, and recently, many advances in the expression of therapeutic glycoproteins with mammalian-type sugar chains in yeast have been demonstrated.

Concept

N-glycosylation pathways in human and yeast are shown in Fig. 1. Early steps of N-glycan processing in the endoplasmic reticulum (ER) are highly conserved from yeast to human, involving the assembly of the lipid-linked oligosaccharide, its site-specific transfer onto the protein by oligosaccharyltransferase complex and the trimming of oligosaccharide by α-glucosidase I, II, and ER α-mannosidase. However, glycosyltransferase reactions in the Golgi apparatus are completely different between yeast and human; mannose residues are transferred by many mannosyltransferases in yeast, whereas in human, mannose residues are first trimmed by Golgi α-mannosidase I, and the following modifications with N-acetylglucosamine, galactose, and sialic acid occur in several glycosyltransferases. In 1992, we isolated and cloned *Saccharomyces cerevisiae* α-1,6-annosyltransferase gene (*OCH1*), which transfers an initial mannose to the ER core high-mannose structure (Nakayama et al. 1992). We also cloned several key genes encoding terminal α-1,3-mannosyltransferase (*MNN1*), mannosylphosphate transferase (*MNN6*) and the positive regulator of Mnn6p (*MNN4*), which are related to make the yeast-specific sugar chains. Further, the disruption of three genes, *OCH1*, *MNN1*, and *MNN4*, resulted in the production of $Man_8GlcNAc_2$ and $Man_5GlcNAc_2$ structures in yeast, which are intermediates of complex-type sugar chains (Chiba et al. 1998). So far, all successful

[1] Research Institute for Cell Engineering, National Institute of Advanced Industrial Science and Technology, Tsukuba Central 6, 1-1-1 Higashi, Tsukuba 305-8566, Japan
Phone: +81-29-861-6160, Fax: +81-29-861-6161
E-mail: jigami.yoshi@aist.go.jp

[2] Research Center for Medical Glycoscience, National Institute of Advanced Industrial Science and Technology, Tsukuba Central 6, 1-1-1 Higashi, Tsukuba 305-8566, Japan

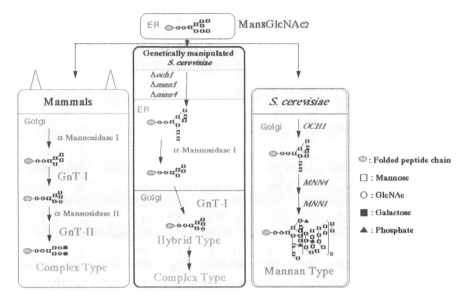

Fig. 1 Schematic representation of *N*-glycosylation pathways in human and yeast, and the strategy for manipulating sugar chains from yeast type to human type

efforts to make mammalian-type *N*-glycosylation pathways in yeast have focused on the deletion of specific yeast genes involved in hypermannosylation, and the introduction of genes responsible for the sugar–nucleotide synthesis, its transport from cytosol to Golgi lumen, and transfer and hydrolysis of sugars. Gerngross' group at Dartmouth College and GlycoFi Inc. reported the complex-type sugar chain production in the methylotrophic yeast *Pichia pastoris* (Hamilton et al. 2003). They also reported the production of a novel glycoprotein, such as erythropoietin, containing a sialylated biantennary sugar chain in another *Pichia* strain (Hamilton et al. 2006). Manipulation of other yeasts, including *Schizosaccharomyces pombe*, *Hansenula polymorpha*, and *Ogataea minuta*, are being attempted to produce humanized oligosaccharides by several other groups.

Because *S. cerevisiae* produces phosphorylated sugar chains as described above, we constructed a new strain with a disruption of both the *OCH1* and *MNN1* genes to produce recombinant lysosomal enzymes for enzyme replacement therapy, in which the responsible enzymes are supplied to the patients of lysosomal diseases (Chiba et al. 2003; also see the Chapter by K. Itoh in *Experimental Glycoscience: Glycobiology*). Since the terminal mannose residues attached to mannose-6-phosphate residue by the phosphodiester linkage in *S. cerevisiae* should be removed for recognition by the human mannose-6-phosphate receptor, we screened for a new bacterial strain that produces an effective α-mannosidase to digest "covered" mannose residue on the glycoprotein. The recombinant α-galactosidase A, which was uncovered by α-mannosidase treatment, was incorporated efficiently into Fabry fibroblasts in culture, and the incorporated enzyme effectively degraded the accumulated ceramide trihexoside (CTH) after 5 days. A biodistribution study in Fabry model mice revealed the partial degradation of the accumulated CTH in some organs after 4 times of weekly injection. These results indicated that the recombinant lysosomal enzyme produced in yeast is useful for providing enzymes more affordably than the current technology.

In contrast, the engineering of *O*-glycosylation has not been attempted in yeast because *O*-mannosylation is one of the specific modifications in yeasts and is vital for yeast cells. However, recently, our group has successfully produced *O*-fucosylated peptide for epidermal growth factor like domain in *S. cerevisiae* by introducing the *Arabidopsis thaliana* genes encoding enzymes converting GDP-mannose to GDP-fucose and human gene for protein *O*-fucosyltranferase I, which will be useful for analyzing the function of *O*-fucosylated proteins, such as the Notch proteins in Notch signaling pathways (Chigira et al. 2008). Mucin-type sugar chain is one of the typical *O*-linked sugar chains in mammals, and it was shown that mucin-type glycoproteins are involved in a variety of biological processes. We have also succeeded in making both *O*-GalNAc peptide and mucin-type glycoprotein in yeast by introducing three responsible genes encoding *Bacillus* UDP-GlcNAc 4-epimerase, human UDP-Gal/GalNAc transporter 2, and human ppGalNAc-T1 (Amano et al. 2008). Based on the development of yeast system to produce mammalian *N*- and *O*-glycosylated sugar chains, it is reasonable to say that the production of therapeutic glycoproteins and glycopeptides by yeast has now become a good candidate in any manufacturing process.

Procedure

Extraction of Yeast Mannoprotein for Glycan Analysis

1. Inoculate 5 ml YPAD medium (2% polypeptone, 1% yeast extract, 2% glucose, 40 mg/L adenine hemisulfate) with a single yeast colony. Grow overnight at 30°C on a shaking incubator.
2. Transfer 1 ml culture to 100 ml YPAD medium.
3. Grow 16–24 h at 30°C on a shaking incubator.
4. Harvest the cells by centrifugation 10 min at 3,000 rpm, 4°C.
5. Wash the cell with distilled water and centrifuge again.
6. Measure the wet weight of the cells.
7. Add 4 ml of 100 mM citrate buffer (pH 7.0) per 1 g pellet.
8. Autoclave the cells at 121°C for 2 h.
9. Recover the supernatant by centrifugation 10 min at 8,000 rpm.
10. Add 3 volumes of cold ethanol and keep it on ice for 30 min.
11. Centrifuge 10 min at 6,500 rpm, 4°C, and recover the pellet.
12. Dry up the pellet, or dissolve the pellet against DW and freeze dry.

Comment

This crude mannoprotein can be applied to PNGaseF treatment or anhydrous hydrazinolysis; however, PNGaseF treatment is recommended for *N*-glycan analysis.

References

Amano K, Chiba Y, Kasahara Y, Kato Y, Kaneko M-K, Kuno A, Ito H, Kobayashi K, Hirabayashi J, Jigami Y, Narimatsu H (2008) Engineering of mucin-type human glycoprotein in yeast cells. Proc Nat Acad Sci USA 105:3232–3237
Chiba Y, Suzuki M, Yoshida S, Yoshida A, Ikenaga H, Takeuchi M, Jigami Y, Ichishima E (1998) Production of human compatible high mannose-type (Man$_5$GlcNAc$_2$) sugar chains in *Saccharomyces cerevisiae*. J Biol Chem 273:26298–26304

Chiba Y, Sakuraba H, Kotani M, Kase R, Kobayashi K, Takeuchi M, Ogasawara S, Maruyama Y, Naka-jima T, Takaoka Y, Jigami Y (2003) Production in yeast of alpha-galactosidase A, a lysosomal enzyme applicable to enzyme replacement therapy for Fabry disease. Glycobiology 12:821–828

Chigira Y, Oka T, Okajima T, Jigami Y (2008) Engineering of a mammalian O-glycosylation pathway in the yeast *Saccharomyces cerevisise*: production of O-fucosylated epidermal growth factor domains Glycobiology, in press, doi:10.1093/glycob/cwn008

Hamilton SR, Bobrowicz P, Bobrowicz B, Davidson RC, Li H, Mitchell T, Nett JH, Rausch S, Stadheim TA, Wischnewski H, Wildt S, Gerngross TU (2003) Production of complex human glycoproteins in yeast. Science 301:1244–1246

Hamilton SR, Davidson RC, Sethuraman N, Nett JH, Jiang Y, Rios S, Bobrowicz P, Stadheim TA, Li H, Choi BK, Hopkins D, Wischnewski H, Roser J, Mitchell T, Strawbridge RR, Hoopes J, Wildt S, Gerngross TU (2006) Humanization of yeast to produce complex terminally sialylated glycoproteins. Science 313:1441–1443

Nakayama K, Nagasu T, Shimma Y, Kuromitsu J, Jigami Y (1992) *OCH1* encodes a novel membrane bound mannosyltransferase: outer chain elongation of asparagine-linked oligosaccharides. EMBO J 11:2511–2519

Solid-Phase Synthesis of Glycopeptides

Yoshiaki Nakahara[1], Hironobu Hojo[2]

Introduction

By considering the inaccessibility of a homogeneous sample from natural sources, a synthetic approach to glycoproteins has been tackled in this laboratory (Nakahara 2003; Hojo and Nakahara 2007). While most of the glycoproteins are too large and complex to be synthesized by a merely chemical procedure, a 20–30 kDa or smaller molecule will be accessible by a combination of solid-phase (glyco)peptide synthesis and segment condensation. Scheme 1 outlines our strategy, which involves (a) synthesis of glycoamino acid building blocks, (b) Fmoc-based solid-phase synthesis, (c) detachment of glycopeptide from the resin synchronized with deprotection of glycan and side-chain functional groups, and (d) segment condensation. In the solid-phase synthesis, the hydroxyl groups of O- or N-glycan are usually masked just as the side-chain functional groups of amino acid to achieve the highest efficiency in every chemical step. On the other hand, minimum protection is desirable in the segment condensation to overcome the poor solubility of the protected (glyco)oligopeptides.

Synthesis of the Glycoamino Acid Building Blocks

A convergent procedure is preferred to efficiently prepare complex glycoamino acids carrying the Fmoc-Asn-OH, Fmoc-Ser-OH, or Fmoc-Thr-OH moiety, which can be used for the solid-phase synthesis along with other amino acids. Employing the reactive benzyl-protected intermediates facilitates such glycoamino acid synthesis. Figure 1 shows the glycoamino acid derivatives, **1–17**, which have been successfully utilized for the synthesis of glycopeptides in this laboratory.

Solid-Phase Synthesis

Solid-phase synthesis is performed with commercial resins with a suitable linker for peptide synthesis. The glycopeptide with either C-terminal carboxylic acid or C-terminal carboxamide is synthesized depending upon the choice of linker. Such commercial linker-fixed resins as HMP-resin, HMPB-BHA-resin, CLEAR acid-resin, Sieber amide-resin, Rink amide MBHA-resin, and CLEAR amide-resin have been used for the glycopeptide synthesis. The Fmoc amino acid-preloaded resins are also commercially available. The solid-phase synthesis is performed by a combination of manual operation with a vortex mixer to introduce glycoamino acid and an automated machine procedure to assemble other amino acids, where DCC-HOBt or HBTU-HOBt is popularly used with a tertiary amine in NMP or DMF to activate the (glyco)amino acid, and piperidine is used to

Department of Applied Biochemistry, Institute of Glycotechnology, Tokai University, 1117 Kitakaname, Hiratsuka, Kanagawa 259-1292, Japan
Phone: +81-463-50-2075, Fax: +81-463-50-2075
E-mail: [1]yonak@keyaki.cc.u-tokai.ac.jp, [2]hojo@keyaki.cc.u-tokai.ac.jp

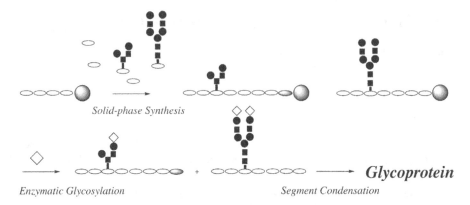

Scheme 1 Synthetic strategy for glycoprotein

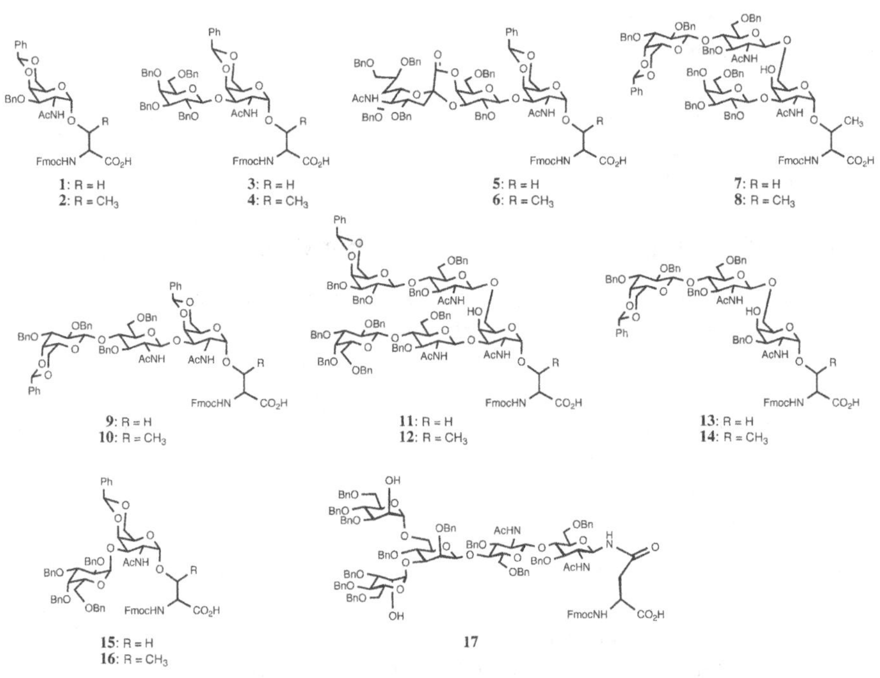

Fig. 1 Glycoamino acid building blocks for the solid-phase glycopeptide synthesis

remove the Fmoc group prior to each coupling reaction. In order to save the amount of a valuable glycoamino acid building block and to facilitate the coupling reaction, modified conditions with an elevated temperature, elongation of the reaction time, and/or a superior coupling reagent such as HATU are often needed. After completion of the solid-phase assembly of the building blocks, the synthesized glycopeptide is split from the resin support by treatment with reagent K (TFA/phenol/water/thioanisole/ethanedithiol, 33:2:2:2:1) for 1 h at ambient temperature. Under the conditions, the benzyl groups in the glycan moiety are removed in part. Complete debenzylation is promoted by low-acidity TfOH (DMS/TFA/m-cresol/TfOH, 3:5:1:1) at −10°C for 2 h to afford the

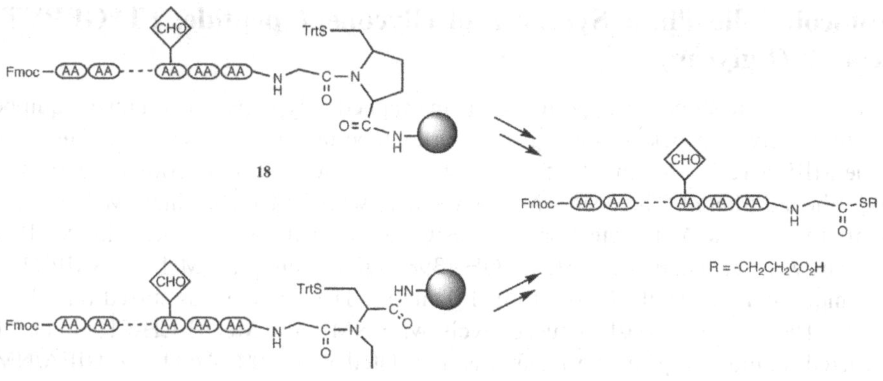

Scheme 2 A new synthesis of glycopeptide thioester by *N*-alkylcysteine devices

desired glycopeptide with minimum scission of acid-labile glycosidic linkages. The crude glycopeptide mixture is precipitated from ether and purified by reversed-phase HPLC. When the synthetic glycopeptide is used as an *N*-terminal component of segment condensation, the glycopeptide is designed in the form of peptide thioester to be selectively activated. Fmoc-based synthesis of the peptide thioester has been permitted under a careful selection of the mildly basic *N*-deprotective agent. However, lowered recovery of the desired product from the resin is ascribed to scission of the thioester linkage most probably caused by the ready diketopiperadine formation at the *N*-deprotected dipeptide stage. This obstacle can be largely overcome by smooth tripeptide preparation when *N*-triisopropylsilyloxycarbonylated dipeptide thioester-resin is *N*-deprotected by a fluoride ion in the presence of the third Fmoc amino acid fluoride (Hojo et al. 2003). Recently, we have proposed a more efficient method for the (glyco)peptide thioester synthesis by utilizing the newly designed devices, **18** and **19**. After cleavage of the synthetic glycopeptides from the resin followed by deprotection of the glycans and the side-chain functional groups, the *N*-alkyl cysteine-type auxiliaries readily convert the glycopeptide to the corresponding thioester via N → S acyl migration and thioester exchange in the presence of 3-mercaptopropionic acid with or without microwave irradiation (Nagaike et al. 2006; Hojo et al. 2007). The glycan structure of synthetic glycopeptide can be further modified by enzymatic glycosylation even if the glycopeptide carries an unstable thioester functionality (Takano et al. 2004).

Segment Condensation

Synthesis of the large glycopeptide is realized by segment condensation, because the standard solid-phase method results in a significant lowering of purity and thus in the overall yield for a product with more than 50 amino acid residues. The glycopeptides with 20–30 residues are coupled through the silver salt-assisted selective activation of the peptide thioester. Repetitive segment condensation has now made it possible to synthesize a 23 kDa glycoprotein carrying 42 *O*-GalNAc residues on the 141 amino acids framework (Hojo et al. 2005). Native chemical ligation is an alternative method to approach the glycoprotein synthesis, although a cysteine residue is necessary at the ligation site (Dawson et al. 1994; Brik et al. 2006).

Protocol: Solid-Phase Synthesis of Glycoheptapeptide, TT*GPPVT (*core 2 *O*-glycan)

All solid-phase reactions were performed in the capped polypropylene test tubes equipped with three-way stopcock by stirring on a vortex tube-mixer. Commercial *N*-Fmoc Rink amide MBHA resin (342 mg, 0.25 mmol) was stirred with 20% piperidine/NMP (3 ml) for 5 min and filtered. The resin was again treated with 20% piperidine/NMP (3 ml) for 15 min to complete *N*-deprotection. The resin was washed several times with NMP. To the resin were added Fmoc-Thr(Bu^t)-OH (398 mg, 1.0 mmol), 1 M DCC/NMP (1 ml, 1.0 mmol), and 1 M HOBt/NMP (1 ml, 1.0 mmol). The mixture was stirred for 1 h and filtered. The resin was washed successively with NMP and MeOH-CH$_2$Cl$_2$ (1 : 1). The unreacted amino group on the resin was acetylated with 10% Ac$_2$O–5% DIEA/NMP (3 ml) for 5 min. After washing with NMP, the *N*-Fmoc group was removed with piperidine, and the second amino acid building block, Fmoc-Val-OH (1.0 mmol), was condensed. Similarly, two proline and glycine residues were sequentially introduced. A part of the *N*-deprotected pentapeptide-resin (58 mg, 32 μmol) was reacted with **8** (118 mg, 63 μmol) in the presence of HATU (23 mg, 60 μmol) and DIEA (17 ml, 95 μmol) in NMP (0.5 ml) at 50°C for 4 h. After washing with NMP and MeOH–CH$_2$Cl$_2$, the resin was *N*-deprotected, reacted with Fmoc-Thr(Bu^t)-OH (50 mg, 125 μmol), and finally *N*-deprotected as described above. A quarter of the resulting resin (30 mg, 8.0 μmol) was used for the experiment to isolate the synthesized glycopeptide by stirring with reagent K [TFA-phenol-deionized water-thioanisole-ethanedithiol (82.5 : 5 : 5 : 5 : 2.5), 0.7 ml] for 1 h. The volatile components in the mixture were removed by blowing nitrogen. Ether was added to the residue, the mixture was centrifuged, and the ethereal layer was decanted. The precipitate was again washed with ether and centrifuged. To the precipitated mixture containing the resin residue was added a mixture of dimethylsulfide (180 μl), m-cresol (60 μl), and TFA (300 μl). The mixture was cooled at −15°C with magnetic stirring. TfOH (60 μl) was added to the stirred mixture, and the mixture was stirred for 2 h before quenching with cooled (−80°C) solution of pyridine (120 μl) in ether (1 ml). The mixture was vigorously stirred for 1 min with a vortex mixer. The precipitate was separated by centrifugation, the ethereal layer was decanted, and the residual precipitate was washed two times with ether as above. The crude product was dissolved in 30% aq. CH$_3$CN and chromatographed by HPLC using a preparative column of RP-18 (250 × 10 mm) under a gradient elution of 7–15% CH$_3$CN (0–16 min, flow rate: 2.5 ml/min) to afford the title glycopeptide (5.6 μmol, 70%). The yield was estimated based on the data of amino acid analysis using the hydrolyzed sample derived from a part of the diluted solution of whole isolated glycopeptide.

References

Brik A, Ficht S, Wong C-H (2006) Strategies for the preparation of homogeneous glycoproteins. Curr Opin Chem Biol 10:638–644

Dawson PE, Muir TW, Clarklewis I, Kent SBH (1994) Synthesis of proteins by native chemical ligation. Science 266:776–779

Hojo H, Nakahara Y (2007) Recent progress in the field of glycopeptide synthesis. Biopolymers (peptide science) 88:308–324

Hojo H, Haginoya E, Matsumoto Y, Nakahara Y, Nabeshima K, Toole BP, Watanabe Y (2003) The first synthesis of peptide thioester carrying N-linked core pentasaccharide through modified Fmoc thioester preparation: synthesis of an N-glycosylated Ig domain of emmprin. Tetrahedron Lett 44:2961–2964

Hojo H, Matsumoto Y, Nakahara Y, Ito E, Suzuki Y, Suzuki M, Suzuki A, Nakahara Y (2005) Chemical synthesis of 23 kDa glycoprotein by repetitive segment condensation: a synthesis of MUC2 basal motif carrying multiple O-GalNAc moieties. J Am Chem Soc 127:13720–13725

Hojo H, Onuma Y, Akimoto Y, Nakahara Y, Nakahara Y (2007) N-Alkyl cysteine-assisted thioesterification of peptides. Tetrahedron Lett 48:25–28

Nagaike F, Onuma Y, Kanazawa C, Hojo H, Ueki A, Nakahara Y, Nakahara Y (2006) Efficient microwave-assisted tandem N- to S-acyl transfer and thioester exchange for the preparation of a glycosylated peptide thioester. Org Lett 8:4465–4468

Nakahara Y (2003) Problems and progress in glycopeptide synthesis. Trends Glycosci Glycotechnol 15:257–273

Takano Y, Hojo H, Kojima N, Nakahara Y (2004) Synthesis of a mimic for the heterogeneous surface of core 2 sialoglycan-linked glycoprotein. Org Lett 6:3135–3138

Efficient Synthesis of Oligosaccharides and Synthesis of Pathogen-Associated Molecular Patterns for Their Biofunctional Studies

Koichi Fukase, Yukari Fujimoto, Katsunori Tanaka

Introduction

Many natural oligosaccharides and glycoconjugates exist in heterogeneous mixtures that are difficult to isolate. Even if the purification is possible, only small amounts are obtained in general. The recent development of oligosaccharide and glycoconjugate synthesis enables the preparation of complex natural oligosaccharides. Synthetic oligosaccharides and glycoconjugates are now used to study their roles in various biologically important processes. In this section, preparation of bacterial glycoconjugate libraries and their use in biofunctional studies are described.

Concept

Library approach based on chemical synthesis has been employed for elucidation of oligosaccharide functions. In this section, we describe the synthesis of libraries of bacterial glycoconjugates such as lipopolysaccharide (LPS) and peptidoglycan (PGN), which have been known as potent immunostimulants, for investigation of their biological functions. Microbial components such as LPS, PGN, double stranded RNA, and bacterial DNA, which do not exist in host animals, are called pathogen-associated molecular patterns (PAMPs). Various immunocompetent cells express pattern recognition receptors (PRRs) that recognize PAMPs. The well-known PRRs are Toll-like receptors (TLRs) and Nod-like receptors (NLRs). Stimulation of TLRs and NLRs by PAMPs activates proinflammatory signaling pathways, which lead to a secretion of various kinds of cytokines that direct inflammation and activation of adaptive immunity. This self-defense system called innate immunity is the first line of the host defense against microorganisms. Since natural PAMPs are usually contaminated with other PAMPs, the precise action mechanisms of innate immunity have been elucidated by using structural definite synthetic specimens.

The LPS is a cell surface glycoconjugate of Gram-negative bacteria and is known as "Endotoxin". The active principle of LPS is its terminal glycolipid moiety lipid A. We have established the efficient synthesis of lipid A and LPS partial structures composed of lipid A and 3-deoxy-D-manno-2-octurosonic acid (Kdo) (Kusumoto and Fukase 2006). We have then prepared the library of lipid A and analogues to elucidate the structural requirements for the immunostimulative and the antagonistic activity with the diversity on the fatty acids and the acidic groups of lipid A.

Department of Chemistry, Graduate School of Science, Osaka University, Machikaneyama-cho 1-1, Toyonaka, Osaka 560-0043, Japan
Phone: +81-6-6850-5388, Fax: +81-6-6850-5419
E-mail: koichi@chem.sci.osaka-u.ac.jp

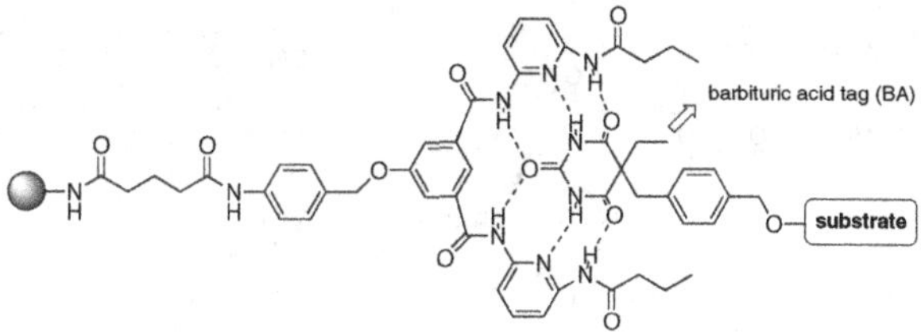

Fig. 1 SAS using interaction between polymer-supported receptor with barbituric acid tag

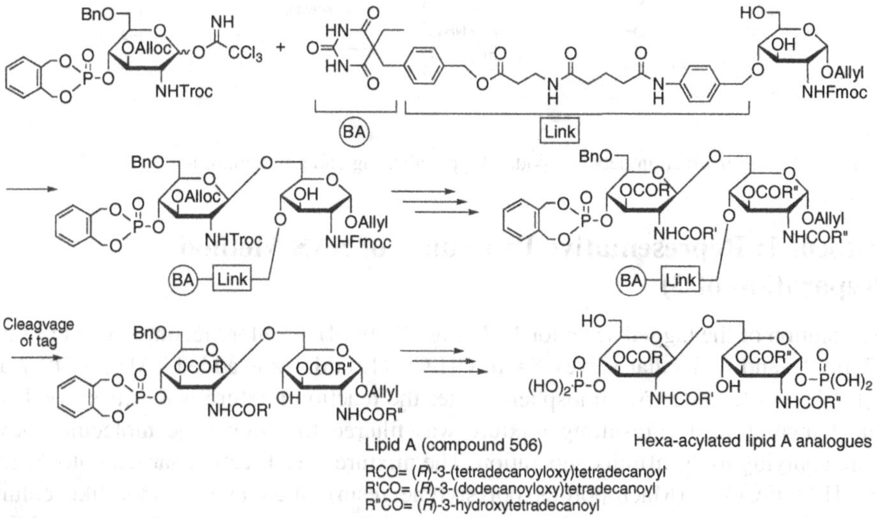

barbituric acid tag (BA)

substrate

Scheme 1 Synthesis of lipid A analogues by using a barbituric acid tag for affinity separation

Lipid A (compound 506)

Hexa-acylated lipid A analogues

RCO= (R)-3-(tetradecanoyloxy)tetradecanoyl
R'CO= (R)-3-(dodecanoyloxy)tetradecanoyl
R"CO= (R)-3-hydroxytetradecanoyl

We prepared lipid A analogues, which have hexaacyl groups with different acylation patterns, by using a barbituric acid tag for affinity separation, based on our "synthesis and affinity separation (SAS)" method (Fig. 1) (Fukase et al. 2005a). This method allowed the efficient synthesis of a focused library of lipid A analogues to reveal the importance of acyl part of lipid A on their biological activities (Scheme 1). In SAS method, the desired tagged compound is separated from the reaction mixture by using specific molecular recognition. After each reaction cycle, the reaction mixture was applied to the column loaded with a receptor to the tag. By using nonpolar eluents such as CH_2Cl_2, the tagged compound was selectively adsorbed on the column, whereas other untagged impurities were washed off. Subsequent desorption by CH_2Cl_2–MeOH (1:1) afforded the desired compound with high purity.

We also employed the interaction between a crown ether (32-crown-10) and an ammonium ion for SAS. Podand-type ether with a pseudo-benzo-31-crown-10 structure was then developed, as it was much easier to synthesize than the crown ether tag and found to show high affinity to the ammonium ion on the solid-support (Fig. 2) (Fukase et al. 2005b). The present SAS method has been successfully applied to the oligosaccharides synthesis.

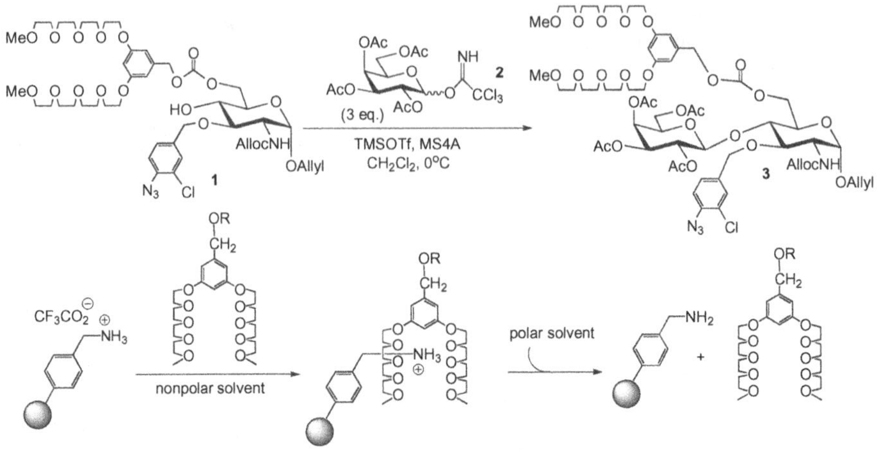

Fig. 2 SAS using interaction between podand-type ether tag and ammonium ion

Protocol 1: Representative Procedure of SAS Method (Preparation of 3)

To a solution of the tagged acceptor **1** (77 mg, 76 μmol), trichloroacetimidate **2** (112 mg, 227 μmol), and molecular sieves 4A in CH_2Cl_2 (1.0 mL) was added TMSOTf (2.7 μL, 15 μmol) at 0°C under Ar atmosphere. After the reaction mixture was stirred for 1 h at room temperature, the resulting mixture was filtered to remove the molecular sieves before applying to the affinity separation. The mixture was directly charged onto ArgoPore-$NH_3^+ \cdot CF_3COO^-$ (Macroporous polystyrene resin) filled in a syringe-like column (Varian, Bond Elut empty cartridges with frits, column size: 2.0 cm × 8.5 cm, resin 3.8 g). The column was prepared as follows. The resin column was washed with CH_2Cl_2–MeOH (1:1) and CH_2Cl_2 (or toluene). The amino groups on the resins were changed to ammonium ions with 10% TFA in CH_2Cl_2 (or toluene), and then excess TFA was washed with CH_2Cl_2 (or toluene). After untagged compounds were washed off with CH_2Cl_2 (200 mL), the tagged product **3** was eluted with CH_2Cl_2–MeOH (1:1, 50 mL). Evaporation of the solvents afforded the desired product **3** as a yellow oil (95 mg, 93%).

We have synthesized PGN partial structures to investigate their action mechanism in immunostimulation. By using efficient synthetic strategy, mono-, di-, tetra-, and octasaccharide fragments of PGN were synthesized in good yields as shown in Scheme 2 (Inamura et al. 2006). A key disaccharide **6** was synthesized by stereoselective glycosylation of a trichloroacetimidate donor **4** with a muramic acid (MurNAc) acceptor **5** by using neighboring group participation of Troc (2,2,2-trichloroethoxycarbonyl) group. The allyl glycoside in **6** was cleaved via isomerization to 1-propenyl group with H_2-activated [Ir(cod)(MePh$_2$P)$_2$]PF$_6$ followed by treatment with I_2 and H_2O, and the product was converted to trichloroacetimidate **7**. Regioselective ring opening of the 4,6-O-benzylidene group in **6** with BH$_3$·Me$_3$N and BF$_3$·Et$_2$O afforded the disaccharide acceptor **8** in 73% yield. Glycosylation of **8** with **7** gave the tetrasaccharide **9** in 79% yield. The octasaccharide **12** was synthesized from **9** in a similar manner. Introduction of peptides and the deprotection were then carried out to give a library of peptidoglycan partial structures.

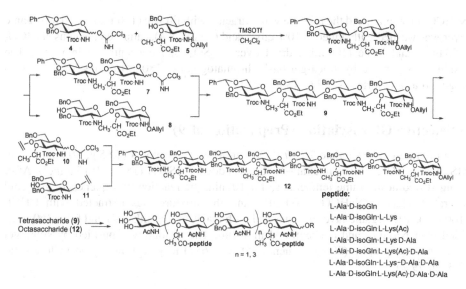

Scheme 2 Synthesis of peptidoglycan partial structures

Protocol 2: Efficient Synthesis of Peptidoglycan Partial Structures, de-allylation of Allyl Glycoside and Formation of Trichloroacetimidate (Preparation of Disaccharide 1-*O*-Trichloroacetimidate 7)

To a degassed solution of the disaccharide allyl glycoside **6** (3.0 g, 2.7 mmol) in dry THF (6 mL) was added [Ir(cod)(MePh$_2$P)$_2$]PF$_6$ (23 mg, 0.027 mmol) activated with H$_2$. After being stirred under nitrogen atmosphere at room temperature for 1 h, iodine (690 mg, 2.7 mmol) and water (10 mL) were added, and the reaction mixture was stirred for additional 10 min. To the reaction mixture was added rapidly aqueous Na$_2$S$_2$O$_3$ (5%, 100 mL). The mixture was then extracted with EtOAc (50 mL). The organic layer was washed with aqueous Na$_2$S$_2$O$_3$ (5%, 50 mL × 2), aqueous sat. NaHCO$_3$ (100 mL × 2), brine (50 mL), dried over Na$_2$SO$_4$, and then concentrated in vacuo. The residue was purified by silica-gel flash chromatography (180 g, toluene-EtOAc = 4 : 1) to give 1-liberated-disaccharide as a pale yellow solid (2.72 g, 93%).

To a solution of 1-OH-disaccharide (2.72 g, 2.56 mmol) in dry CH$_2$Cl$_2$ (6 mL) at rt were added Cs$_2$CO$_3$ (417 mg, 1.28 mmol) and CCl$_3$CN (3.7 mL, 25.6 mmol). After being stirred for 1 h, insoluble materials were filtered off through celite and concentrated. The residue was lyophilized from benzene to give **7** as a pale yellow solid (3.04 g), which was used for subsequent glycosylation without purification.

Reductive Opening of Benzylidene Acetal Using Me$_3$N·BH$_3$ and BF$_3$·OEt$_2$ (Preparation of 4′-*O*-deprotected-disaccharide 8)

To a solution of **6** (1.5 g, 1.36 mmol) and Me$_3$N·BH$_3$ (150 mg, 2.05 mmol) in dry CH$_3$CN (13 mL) at 0°C was added BF$_3$·OEt$_2$ (960 mg, 6.80 mmol) dropwise, and the mixture was stirred at rt for 30 min. The reaction was then quenched with ice and saturated aqueous

NaHCO$_3$ (100 mL), and the mixture was extracted with EtOAc (100 mL × 2). The organic layer was washed with aqueous 10% citric acid (15 mL × 4), saturated aqueous NaHCO$_3$ (150 mL), and brine (100 mL), dried over Na$_2$SO$_4$, and concentrated in vacuo. The residue was purified by silica-gel flash chromatography (180 g, toluene-AcOEt = 4 : 1) to give **8** as a colorless solid (1.13 g, 73%).

β-Selective Glycosylation (Preparation of 9)

To a mixture of the imidate **7** (2.7 g, 38 mmol), the acceptor **8** (1.65 g, 30 mmol), and MS4A in dry CH$_2$Cl$_2$ (75 mL) at −15°C was added TMSOTf (18 μL, 0.15 mmol). After being stirred at the same temperature for 10 min, the reaction was quenched with cold saturated aqueous NaHCO$_3$ (100 mL), and the mixture was extracted with CHCl$_3$ (100 mL). The organic layer was washed with NaHCO$_3$ (60 mL × 2) and brine (60 mL), dried over Na$_2$SO$_4$, and concentrated in vacuo. The residue was purified by silica-gel flash chromatography (300 g, toluene-EtOAc = 6 : 1) to give **9** as a pale yellow solid (2.33 g, 79%).

Final Deprotection by Catalytic Hydrogenation (Preparation of Tetrasaccharide Tripeptide)

To a solution of the protected tetrasaccharide tripeptide (95 mg, 36 μmol) in AcOH (3 mL) was added palladium hydroxide (100 mg) in AcOH and stirred under H$_2$ (20 atm) for 1 day. The Pd catalyst was filtered off by celite, and the filtrate was concentrated. The residue was lyophilized from acetonitrile-H$_2$O to give the free tetrasaccharide tripeptide (39 mg, 50%) as a white powder.

By using these synthetic fragments, we revealed that not only glycan but also peptide was essential for expression of the immunostimulating activity. TLR2, which is one of the candidates of PGN receptor, was not stimulated by the series of synthetic PGN partial structures. On the other hand, intracellular receptor Nod2 recognized the partial structures containing muramyl dipeptide (MDP) moiety. These results indicated that Nod2 is the intracellular receptor of muropeptides derived from peptidoglycan, and the minimal ligand of Nod2 is MDP. Recognition by peptidoglycan recognition proteins (PGRPs) has also been elucidated by using the synthetic fragments. Dziarski et al. reported that human PGRP-L is *N*-acetylmuramoyl-L-alanine amidase, which hydrolyzes the amide bond between MurNAc and L-Ala of bacterial PGN. The minimum PGN fragment hydrolyzed by PGRP-L is MurNAc-tripeptide. Ezekowitz and co-workers have revealed that human PGRP-S inhibits the growth of bacteria. PGRP-S bound to the tetrasaccharide having two tripeptide chains and the tetrasaccharide having two tetrapeptide chains, suggesting that the third amino acid in peptide chain determines the binding affinity to PGRP-S.

Melanin synthesis of the arthropods is essential for defense and development. The melanization cascade is activated by bacterial PGN or fungal β-1,3-glucan. Lee and co-workers reported that the tetrasaccharide tetrapeptide functioned as a competitive inhibitor of the natural PGN-induced melanization reaction (Park et al. 2006). By using a tetrasaccharide tetrapeptide-coupled column, the Tenebrio molitor PGRP (Tm-PGRP) was purified without activation of the prophenoloxidase.

References

Fukase K, Zhang S-Q, Fukase Y, Umesako N, Kusumoto S (2005a) Synthesis based on affinity separation: a new methodology for high throughput synthesis using affinity tags. ACS Symp Ser 892:87–98

Fukase K, Takashina M, Hori Y, Tanaka D, Tanaka K, Kusumoto S (2005b) Oligosaccharide synthesis by affinity separation based on molecular recognition between podand ether and ammonium ion. Synlett 2342–2346

Inamura S, Fujimoto Y, Kawasaki A, Shiokawa Z, Woelk E, Heine H, Lindner B, Inohara N, Kusumoto S, Fukase K (2006) Synthesis of peptidoglycan fragments and evaluation of their biological activity. Org Biomol Chem 4:232–242

Kusumoto S, Fukase K (2006) Synthesis of endotoxic principle of bacterial lipopolysaccharide and its recognition by the innate immune systems of hosts. Chem Rec 6:333–343

Park JW, Je B-R, Piao S, Inamura S, Fujimoto Y, Fukase K, Kusumoto S, Soederhaell K, Ha N-C, Lee BL (2006) A synthetic peptidoglycan fragment as a competitive inhibitor of the melanization cascade. J Biol Chem 281:7747–7755

Sugar Polymers (Dendrimers and Pendant-Type Linear Polymers)

Koji Matsuoka

Introduction

The glycoconjugates that are called the glycoproteins, the glycolipids, and the proteoglycans exist ubiquitously in vivo. The oligosaccharide chains of glycoconjugates are the third chains of biomacromolecules next to DNAs and proteins and have very complex structures. Affinity of a monomeric sugar chain of bioactive glycoconjugates against a variety of proteins such as carbohydrate-binding proteins, lectins, and enzymes, is low, usually in the millimolar range. In the 1970s, Lee reported remarkable enhancement of the binding affinity by means of a multivalent-type sugar substrate, so-called "sugar clustering effect" (Lee et al. 1983). Results of recent studies on the cell surface suggested that glycoprotein and glycolipids form microdomains such as rafts or patches which assemble on their own to produce natural glycoclusters. Thus, the accumulation of sugar chains having weak binding affinity results in the formation of glycoclusters, and the glycoclusters acquire high affinity for biological interactions. In this chapter, we describe concepts of sugar polymers through synthetic studies of glycoclusters including carbosilane dendrimers uniformly functionalized with carbohydrate moieties through covalent bonds and linear polymers having pendant-type carbohydrate moieties.

Carbosilane Dendrimers Functionalized with Carbohydrates

We have selected carbosilane dendrimers as core scaffolds for assembly of functional groups as well as functional molecules. Unique advantages by using carbosilane dendrimers as the core frame are: (1) simplicity of the synthetic process to extend the generation, (2) accessibility to the polymer with definite molecular weight and definite number of terminal functions, (3) neutral nature in contrast to polyamine-type dendrimers, (4) chemical and biochemical stability and (5) biological inertness (Matsuoka et al. 2006a).

We describe here a synthetic assembly of carbohydrate molecules using carbosilane dendrimers as core scaffolds, because carbosilane dendrimers have various merits as mentioned above. Figure 1 shows a series of carbosilane dendrimers (Matsuoka et al. 2006b, c). These dendrimers display unique shapes, generations, and number of functional molecules. We call these dendrimers "SUPER TWIG," and the column indicates the shape of dendrimer cores where the left column shows a fan shape, the center column shows a ball shape, and the right column shows a dumbbell shape. The number in parenthesis is the number of generations, and the number after the parenthesis shows the number of functional groups at the end terminal. Examples of glycodendrimers are summarized in Fig. 2. These glycodendrimers showed excellent biological activities in vitro as well as in vivo experiments.

Area for Molecular Function, Division of Material Science, Graduate School of Science and Engineering, Saitama University, Sakura, Saitama 338-8570, Japan

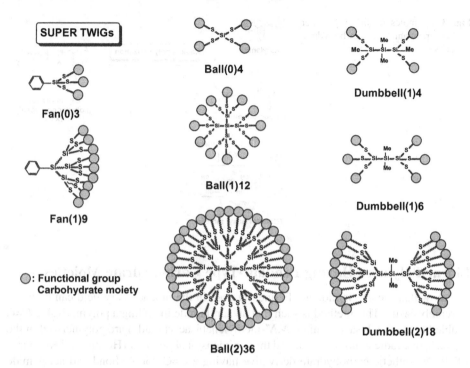

Fig. 1 A series of carbosilane dendrimers uniformly functionalized with functional molecules (SUPER TWIGs)

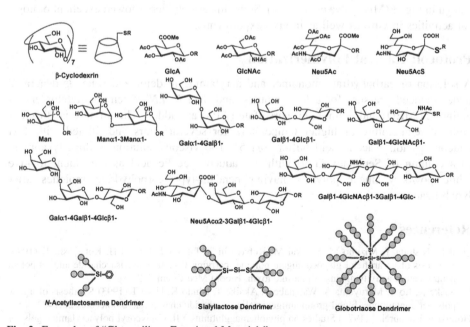

Fig. 2 Examples of "Glyco-silicon Functional Materials"

Fig. 3 Examples of linear polymer having pendant-type carbohydrate moieties

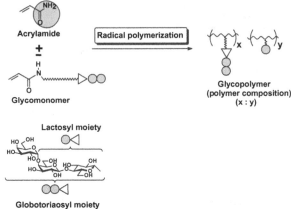

Linear Polymers Having Pendant-Type Carbohydrate Moieties

We know that the acrylamide gel frequently used in the biochemistry field can be made extremely easily. This method is such that an acrylamide including a polymerizable C=C double bond and a bisacrylamide (*N*,*N*′-methylenebisacrylamide) are polymerized in the presence of a radical initiator to afford three-dimensional polymer (Hořejší and Kocourek 1973). A synthetic carbohydrate derivative having a C=C double bond and acrylamide are treated with the same condition to provide water-soluble linear polymer having pendant-type carbohydrate moieties (Nishimura et al. 1990; Hatanaka et al. 1993; Dohi et al. 1999). We have prepared some linear polymers having glycoepitopes, which is shown in Fig. 3 (Miyagawa et al. 2004). Some linear polymers showed excellent biological activities in vitro as well as in vivo experiments.

Protocol: Radical Polymerization

A solution of carbohydrate monomer and acrylamide in deionized water is deaerated under reduced pressure for a few minutes, and then *N*,*N*,*N*′,*N*′-tetramethylethylenediamine (TEMED) and ammonium persulfate (APS) are added. The mixture is stirred at ambient temperature or higher temperature for several hours and diluted with 1 M aqueous pyridine–acetic acid buffer (pH 5). The viscous solution is directly applied to a column of Sephadex G-50 with 5% aqueous acetic acid as the eluent to give white powdery linear polymer having pendant-type carbohydrate moieties after lyophilization.

References

Dohi H, Nishida Y, Mizuno M, Shinkai M, Kobayashi T, Takeda T, Uzawa H, Kobayashi K (1999) Synthesis of an artificial glycoconjugate polymer carrying P^k-antigenic trisaccharide and its potent neutralization activity against Shiga-like toxin. Bioorg Med Chem 7:2053

Hatanaka K, Ito Y, Maruyama A, Watanabe Y, Akaike T, Ishido K, Uryu T (1993) Synthesis of a new polymer containing a blood-group antigenic oligosaccharide chain. Macromolecules 26:1483

Hořejší V, Kocourek J (1973) Studies on phytohemagglutinins XII. *O*-glycosyl polyacrylamide gels for affinity chromatography of phytohemagglutinins. Biochim Biophys Acta 297:346

Lee YC, Townsend RR, Hardy MR, Lönngren J, Arnarp J, Haraldsson M, Lönn H (1983) Binding of synthetic oligosaccharides to the hepatic Gal/GalNAc lectin. J Biol Chem 258:199

Matsuoka K, Hatano K, Terunuma D (2006a) Glycodendrimers using carbosilanes as core scaffolds. Nanotechnology in carbohydrate chemistry. In: Yuasa H (ed) Transworld Research Network, p 89

Matsuoka K, Terabatake M, Umino A, Esumi Y, Hatano K, Terunuma D, Kuzuhara H (2006b) Carbosilane dendrimers bearing globotriaoses: syntheses of globotrioasyl derivative and introduction into carbosilane dendrimers. Biomacromolecules 7:2274

Matsuoka K, Terabatake M, Esumi Y, Hatano K, Terunuma D, Kuzuhara H (2006c) Carbosilane dendrimers bearing globotriaoses: construction of a series of carbosilane dendrimers bearing globotriaoses. 7:2284(ibid)

Miyagawa A, Kurosawa H, Watanabe T, Koyama T, Terunuma D, Matsuoka K (2004) Synthesis of glycoconjugate polymer carrying globotriaose as artificial multivalent ligand for shiga toxin-producing *Escherichia coli* O157: H7. Carbohydr Polym 51:441–450

Nishimura S-I, Matsuoka K, Kurita K (1990) Synthetic glycoconjugates: simple and potential glycoprotein models containing pendant *N*-acetyl-D-glucosamine and *N,N'*-diacetylchitobiose. Macromolecules 23:4182

Rapid Synthesis of Oligosaccharides: Resin Capture–Release Strategy

Yukishige Ito, Shinya Hanashima

Introduction

A number of methodologies for efficient and selective O-glycoside bond formation have been successfully developed. Using these methods, syntheses of complex oligosaccharides have been achieved. However, chemical synthesis of glycoprotein-derived oligosaccharide or their derivative is technically demanding and highly time-consuming. In order that synthetic chemistry is going to function as the driving force in glycobiology, speeding up and automation are the key issues (Seeberger and Haase 2000).

Solution-Phase Oligosaccharide Synthesis with Real-Time Monitoring

Without doubt, solid-phase synthesis is the key technology for the development of automated oligopeptide or oligonucleic acid synthesizers. Establishment of a similar technology seems to be ideal for facilitating the synthesis of oligosaccharides. Our approach features the use of low-molecular weight polyethylene glycol (LPEG) as a soluble support (Ando et al. 2001a). It was expected that the reduction of reactivity frequently observed in solid-phase synthesis can be minimized by using LPEG, because (1) all reactions can be conducted in a homogeneous solution and (2) the size of the polymer support is minimum (M.W. ~550). In addition, the identity and purity of products can be verified easily by spectroscopic means, such as ^{13}C and ^{1}H NMR. Unlike the case of solid-support synthesis, large excess of donor may not be necessary, because the reactivity of acceptor would be comparable with the non-supported one. Furthermore, straightforward color tests are available to monitor the removal of temporary protecting group (Ando et al. 2001a).

However, there is no guarantee that all glycosylations proceed into completion, and the accumulation of incompletely elongated products may not be avoidable. Addressing this particular problem, we have developed the "Resin Capture-Release" strategy, which enables the selective isolation of desired oligosaccharides (Ando et al. 2001b). It was successfully applied to the synthesis of sialylated complex-type glycans (Hanashima et al. 2005) and bisubstrate-type inhibitors of N-acetylglucosaminyl transferases (GnT-V and GnT-IX) (Hanashima et al. 2006).

The blueprint of our strategy is depicted in Scheme 1 (Hanashima et al. 2003). Thus, starting with an acceptor 1^n that was supported on LPEG, chain elongation was conducted

RIKEN (The Institute of Physical and Chemical Research), 2-1 Hirosawa, Wako, Saitama 351-0198, Japan;
CREST, JST, Kawaguchi, saitama 332-1102, Japan
Phone: +81-48-467-9430, Fax: +81-48-462-4680
E-mail: yukito@riken.jp

Scheme 1 Concept of the capture-release strategy

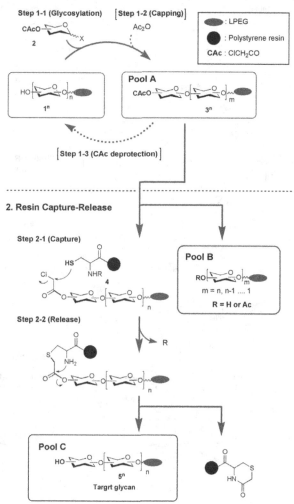

by glycosylation with a glycosyl donor **2**, which carries a chloroacetyl (CAc) group as a temporary protecting group (**Step 1-1**). Taking advantage of the polarity of LPEG, impurities derived from the donor are removed by a short silica gel column. Namely, after elution with EtOAc, the LPEG tagged components containing glycosyleted product 3^n (m = n) are then eluted with more polar solvent systems such as EtOAc/MeOH. Treatment with resin-supported Cys **4** specifically captures 3^n, through chemoselective reaction between thiol and CAc group (**Step 2-1**). Subsequent removal of the amine protection triggers cyclization (**Step 2-2**) to release the target glycan 5^n.

The overall efficiency of this process could be further enhanced, when **Step 1-2** (capping of the remaining hydroxyl group) and **Step 1-3** (partial deprotection of the CAc group to generate the acceptor 1^n; n = m + 1) are incorporated. In that case, capture-release purification is required only after final glycosylation. Repetition of the cycle

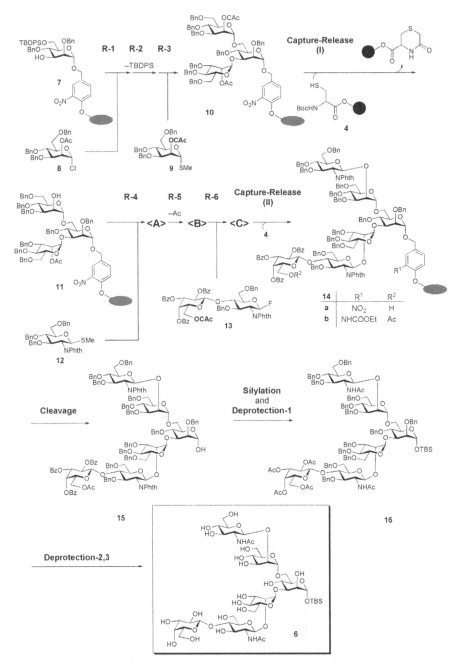

Scheme 2 Synthesis of branched hexasaccharide on soluble support

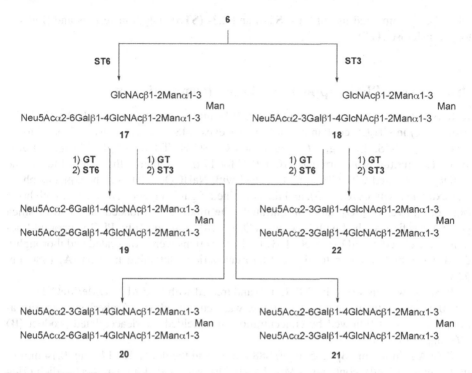

Scheme 3 Diversification of the common precursor by enzymatic glycosylations

(**Steps 1-1, 1-2,** and **1-3**) produces **Pool A**, in which successfully glycosylated product 3^n is the only compound carrying a CAc group. After the capture step, deletion products (**Pool B**) can be removed simply by filtration and washing.

As an example, divergent synthesis of sialylated complex-type glycans was conducted (Hanashima et al. 2005). It consists of (1) the chemical synthesis of the hexasaccharide **6** as a common intermediate (Scheme 2) and (2) enzyme-mediated introduction of sialic acid (NeuAc) and galactose (Gal) residues (Scheme 3). Synthesis of the hexasaccharide **6** was commenced with LPEG-supported acceptor **7**. It was glycosylated successively with mannosyl donors **8** and **9** to provide **10** (**R-1 ~ 3**), which was subjected to capture-release purification using Boc-Cys-resin **4**, to give **11**. Further conversion to **14a** was carried out by successive glycosylations with **12** and **13** (**R-4 ~ 6**), which was followed by the 2nd capture-release purification with **4**. Liberation of the oligosaccharide from LPEG was carried out in three steps; reduction of the nitro group and ethoxycarbamate formation gave **14b**, which was treated with 15% TFA to give hemiacetal **15**.

After being converted to the *t*-butyldimethylsilyl (TBS) glycoside by using *N,O*-bis-t-butyldimethylsilylacetamide and 0.05 equivalent of tetrabutylammonium fluoride (TBAF) (**Silylation**), deprotection was conducted in three steps via **16** (**Deprotection-1 ~ 3**).

Hexasaccharide **6** was subjected to a series of enzymatic glycosylations. It could be diversified to regioisomeric Neu5Ac$_2$Gal$_2$GlcNAc$_2$Man$_3$ structures, **19** (2,6/2,6), **20** (2,3/2,6), **21** (2,6/2,3), **22** (2,3/2,3) through monosialylated heptasaccharides **17** (2,6) and

18 (2,3), by combined use of 2,6- (**ST6**) and 2,3- (**ST3**) sialyltransferases and β-galac-tosyltransferase (**GT**).

Protocol-1: LPEG-support Synthesis (Preparation of 14)

R-4: A solution of acceptor **11** (432 mg, 0.19 mmol) and thioglycoside **12** (241 mg, 0.40 mmol) in CH_2Cl_2 containing molecular sieves (MS) 4A (1.3 g) was added N-iodo-succinimide (NIS, 136 mg, 0.60 mmol) and CF_3SO_3H (TfOH, 10 µl, 0.11 mmol) at −10°C. The mixture was stirred at −10 to 0°C for 17 h, and filtered through Celite and the filtrate was diluted with $CHCl_3$ and washed with $NaHCO_3$ solution. The aqueous phase was extracted with $CHCl_3$ (×3), and the combined organic layers were washed with brine (×2), dried over Na_2SO_4, and concentrated. The mixture was loaded on a pad of silica gel and was first washed with Hexane-AcOEt (1:4). Subsequently, PEG-supported frac-tions were eluted by EtOAc-MeOH (3:1). The fractions were evaporated and thoroughly dried under high vacuum to give the tetrasaccharide-containing fraction ⟨**A**⟩ (486 mg, 92%).

R-5: ⟨**A**⟩ was dissolved in THF (3 ml) and treated with 0.05 M NaOMe/MeOH (5 ml) for 10 h at room temperature. The mixture was neutralized with Amberlyst resin. Filtra-tion of resin was followed by concentration that yielded the deacetylated product ⟨**B**⟩ (460 mg, 96%).

R-6: A solution of ⟨**B**⟩ (235 mg, 0.088 mmol) and the donor **13** (146 mg, 0.14 mmol) in toluene (3.5 ml) containing MS 4A (0.8 g) was added to hafnocene dichloride (Cp_2HfCl_2, 54 mg, 0.14 mmol) and silver trifluoromethanesulfonate (AgOTf, 80 mg, 0.31 mmol) at −20°C. The mixture was stirred at −20 to 0°C for 12 h. The mixture was filtered through Celite and the filtrate was diluted with $CHCl_3$ and washed with $NaHCO_3$ solution. The aqueous phase was extracted with $CHCl_3$ (×3), and the combined organic layers were washed with brine (×2), dried over Na_2SO_4, filtered, and concentrated. The mixture was loaded on a pad of silica gel and was first washed with Hexane-AcOEt (1:4) to remove non-PEG supported materials. Subsequently, PEG-supported fractions were eluted by EtOAc-MeOH (3:1) to give the hexasaccharide ⟨**C**⟩ (299 mg, 92%); ^{1}H NMR (400 MHz, $CDCl_3$) δ 8.01-6.76 (90H), 5.79 (dd, 1H, J = 8.0, 10.8 Hz), 5.75 (d, 1H, J = 3.2 Hz), 5.37 (dd, 1H, J = 3.4, 10.2 Hz), 5.24 (d, 1H, J = 7.8 Hz), 4.92-3.47 (overlapped with PEG methylene signal), 3.37 (s, 3H, PEGOMe), 3.33 (d, 1H, J = 10.5 Hz), 3.21 (d, 1H, J = 10.4 Hz), 3.13 (d, 1H, J = 11.2 Hz), 2.87 (dd, 1H, J = 5.6, 10.8 Hz), 2,61 (m, 1H), 2.31 (m, 1H).

Capture-release (II): The hexasaccharide ⟨**C**⟩ (160 mg, 0.044 mmol) in DMF (4.6 ml) was treated with (Boc)Cys-Merrifield resin **4** (0.60 mmol/g, 600 mg, 0.36 mmol) and i-Pr$_2$NEt (700 µl). The mixture was mixed for 18.5 h at 80°C, and washed with $CHCl_3$/MeOH, and dried under reduced pressure. To the resin was added 10% TFA in CH_2Cl_2 (5.6 ml), vortex mixed for 30 min, washed with $CHCl_3$/MeOH, and dried under reduced pressure. Then, 10% piperidine in THF (10 ml) was added and the mixture was shaken for 13 h. The resin was rinsed with $CHCl_3$/MeOH. The rinsing was diluted with $CHCl_3$ and washed with 0.5 M HCl aq. and brine, and dried over Na_2SO_4, filtered, and concen-trated. The mixture was loaded on a pad of silica gel and was first washed with Hexane-AcOEt (1:4) to remove non-PEG supported materials. Subsequently, PEG-supported fractions were eluted by EtOAc-MeOH (3:1) to give the hexasaccharide **14a** (138 mg, 88%).

Protocol-2: Hexasaccharide Isolation and Deprotection

Cleavage from LPEG: A solution of the hexasaccharide **14a** (133 mg, 0.037 mmol) in pyridine (2 ml) was added Ac_2O (1 ml), and stirred for 12 h at room temperature. The mixture was evaporated, co-evaporated with toluene, and dried under vacuum. The residue was dissolved in dichloroethane (1.5 ml)-EtOH (1.5 ml) and Mo $(CO_2)_6$ (100 mg, 0.38 mmol) was added. The mixture was stirred for 4.5 h under reflux, then cooled to room temperature, diluted with $CHCl_3$ and washed with aq. $NaHCO_3$. The aq. phase was extracted with $CHCl_3$ (3×), and the combined organic layers were washed with brine, and dried over Na_2SO_4, concentrated and dried under reduced pressure. The residue was dissolved in CH_2Cl_2, and ethyl chloroformate (40 μl, 0.42 mmol) and i-Pr_2NEt (130 μl, 0.75 mmol) were added. After being stirred for 23 h at room temperature, the mixture was concentrated and co-evaporated with toluene. The mixture was loaded on a pad of silica gel and was first washed with Hexane-AcOEt (1:4) to remove non-PEG-supported materials. Subsequently, PEG-supported fractions were eluted by EtOAc-MeOH (3:1). The fractions were evaporated and thoroughly dried under high vacuum to give **14b**. It was treated with 15% TFA in CH_2Cl_2 (3 ml) for 4 h. The mixture was concentrated, co-evaporated with toluene, and purified with SiO_2 chromatography (tol:EtOAc = 8:1 ~ 2:1) to yield the hemiacetal **15** (90 mg, 88%).

Silylation and deprotection-1: To a solution of the hemiacetal **15** (105 mg, 0.038 mmol) in N-methylpyrolidone (NMP, 0.5 ml) were added N,O-bis (t-butyldimethylsilyl)acetamide (BTBSA: 62 μl, 0.18 mmol) and TBAF (7 μl, 0.5 M in THF). The mixture was stirred for 50 min at room temperature, and quenched with 10% citric acid solution. The mixture was extracted with $CHCl_3$ (3×), and the combined organic layers were washed with brine (2×), dried over Na_2SO_4, concentrated, and dried under reduced pressure. The crude hexasaccharide in 1-BuOH (4 ml) was added ethylenediamine (1 ml), and stirred at 95°C for 19 h. After cooling, solvent was removed, and the residue treated with pyridine (3 ml) and Ac_2O (1.5 ml). The mixture was stirred at 40°C for 16 h and diluted with $CHCl_3$. The solution was washed with 10% citric acid solution and brine, dried over Na_2SO_4, filtered, and concentrated, and purified with preparative TLC (toluene: EtOAc = 2:1) to afford **16** (53 mg, 56%).

Deprotection-2,3: A solution of **16** (55 mg, 0.021 mmol) in 0.05 M NaOMe in MeOH (3 ml) was stirred at 40°C for 24 h. After cooling to room temperature, the mixture was directly loaded on gel filtration chromatography (Sephadex LH-20, $CHCl_3$:MeOH = 1:4) to give the tetraol (47 mg, 92%). It was dissolved up in MeOH (4 ml)-H_2O (1 ml) and hydrogenated over 20% $Pd(OH)_2$/C (79 mg) under atmospheric pressure at r.t. After 24 h, the mixture was stirred under N_2 atmosphere for 30 min. The catalyst was removed by filtration through Celite, and the filtrate was concentrated, and purified by reverse phase SepPak (C-18) cartridge (MeOH:H_2O = 0:100 to 50:50) to afford **6** (22 mg, 93%).

$[\alpha]_D{}^{24}$ + 28 (c 0.29 H_2O); ^1H NMR (500 MHz, D_2O, t-BuOH as an internal standard; 1.24 ppm, 40°C) δ; 5.12 (d, 1H, J = 1.4 Hz), 5.11 (s, 1H), 4.88 (d, 1H, J = 1.4 Hz), 4.58 (d, 1H, J = 8.7 Hz), 4.46 (d, 1H, J = 7.8 Hz), 4.16 (dd, 1H, J = 1.4, 3.7 Hz), 4.08 (dd, 1H, J = 1.4, 3.7 Hz), 3.99–3.42 (m), 2.05 (s, 3H), 2.04 (s, 3H), 0.92 (s, 9H), 0.17 (s, 3H), 0.16 (s, 3H); ^{13}C NMR (125 MHz, D_2O, MeOH as an internal standard, 40°C) δ; 175.1, 103.3, 100.3, 100.0, 99.9, 97.2, 95.1, 79.0, 78.4, 77.5, 76.8, 76.3, 75.7, 75.1, 73.7, 73.6, 73.3, 72.9, 72.3, 71.8, 71.6, 71.3, 70.3, 70.0, 69.9, 68.9, 67.7, 67.5, 66.5, 66.4, 61.9, 61.7,

61.3, 61.0, 60.4, 55.8, 55.4, 25.3, 22.8, 22.7, 17.6, -5.1, -5.9. HR-ESI-MS: Calcd $(C_{46}H_{82}N_2O_{31}SiNa)$ $[M+Na]^+$ = 1209.4569. Found: 1209.4523.

References

Ando H, Manabe S, Nakahara Y, Ito Y (2001a) Tag-reporter strategy for facile oligosaccharide synthesis on polymer support. J Am Chem Soc 123:3848–3849

Ando H, Manabe S, Nakahara Y, Ito Y (2001b) Solid phase capture-release strategy in soluble polymer support oligosaccharide synthesis. Angew Chem Int Ed 40:4725–4728

Hanashima S, Manabe S, Ito Y (2003) Polymer-resin hybrid capture-release strategy for rapid oligosaccharide construction. Synlett 979–982.

Hanashima S, Manabe S, Ito Y (2005) Divergent syntheses of sialylated glycan chains: combined use of polymer support, resin capture-release and chemo-enzymatic strategies. Angew Chem Int Ed 44:4218–4224

Hanashima S, Inamori K-I, Manabe S, Taniguchi N, Ito Y (2006) Systematic synthesis of bisubstrate-type inhibitors of N-acetylglucosaminyltransferases. Chem Eur J 12:3449–3462

Seeberger PH, Haase WC (2000) Solid-phase oligosaccharide synthesis and combinatorial carbohydrate libraries. Chem Rev 100:4349–4394

Glossary

8-aminopyrene-1,3,6-trisulfonate (APTS) It is a fluorescent-labeling group for reducing ends of glycans. This labeling helps to achieve fast and supersensitive detection of glycans by capillary electrophoresis with argon laser-induced fluorescence detection.

α-1,6-fucosyl transferase This enzyme is also called Fut8. It transfers Fucose to the GlcNAc residue at the most proximity of N-linked glycan chain by the linkage of α-1,6. It has been reported to have pivotal roles in the context of development or differentiation.

α-dystroglycanopathy α-dystroglycanopathy collectively means the muscular dystrophy syndromes resulted from glycan structure deficiency of α-dystroglycan. It is proposed that the malformation of the glycans in α-dystroglycan hinders the binding to ligands, such as laminin, and results in muscular dystrophy because of the degeneration of the intra- or intercellular connection.

α-galactose epitope Humans, primates, and old world monkeys do not have the glycan epitope Galα1-3Gal structure, because the α3 galactosyltransferase responsible for the structure became pseudo-gene in those mammals. Human serum contains approximately 1% of natural antibody IgG against Galα1-3Gal, which would give rise to a serious problem when porcine organs are to be subjected to xenograft into humans.

α-galactosyl ceramide See "NKT cells".

α-mannosidase α-mannosidase is one of the hydrolyzing enzymes, which cleaves off the α1-2, α1-3, and/or α1-6 mannose residue. It includes the processing enzymes localized in the endoplasmic reticulum or Golgi structure, and degrading enzymes that function on the glycoproteins located in the lysosome or cytosol.

α-N-acetylglucosamine residue It was discovered with a structure, GlcNAcα1-4Galβ, at the non-reducing end of O-linked glycan. Humans have such a structure, specifically in the mucus secreted by gland mucus cells dwelling in the middle to the lower layer of the gastric mucosa, the Brunner's gland in the duodenal mucosa, or pancreatic duct epithelial cells showing metaplasia into gastric pyloric gland tissue. This structure was also highly expressed in the gastric, pancreatic or biliary cancer. Thus, it is considered that the α-GlcNAc residue is the cancer-related glycoantigens in those cancers.

ABO blood group antigens The ABO antigens, major blood group antigens, were originally identified in erythrocytes, but they are also found on the mucosal epithelial cells or in the mucosa of various tissues. A type antigen was identified to have a glycoantigen structure, GalNAcα1,3Gal(Fucα1,2)-R, whereas B type has Galα1,3Gal(Fucα1,2)-R, and they are found in glycoproteins and glycolipids.

Highly homologous allelic genes at the ABO blood group locus on the chromosome 9 specify the specificity of transferases involved in the ABO antigen synthesis. H antigen (Fucα1,2Gal-R) is the precursor for those structures.

Adhesion molecule Adhesion molecule is involved in cell–cell and cell–matrix interactions. Epithelial cell shows three types of cell junctions: tight junction, adherence junction, and desmosome. A variety of adhesion molecules, claudine, ocludin, or cadherin, are known to participate in cell–cell interactions. Hemidesmosome is known as the adhesion machinery between cell and extracellular matrix, and integrin molecules are involved in the cell-matrix interaction.

Algorithm Algorithm is a type of effective method, which has a definite list of well-defined instructions for completing a task. Program is the instruction for a computer to carry out the task on the basis of the algorithm.

Alzheimer disease/Alzheimer's disease/AD Alzheimer disease (AD) is a neurodegenerative disease and is the most common cause of dementia. AD is characterized by loss of neurons and synapses, which results in gross atrophy of the affected brain regions. Typical pathological features of AD are deposition of β-amyloid peptide (Aβ) and neurofibrillary tangles in the brain. Deposition of Aβ in the brain is considered to be a major cause of AD, and therapeutic agents to inhibit Aβ production or to stimulate Aβ clearance are in a developmental stage.

Animal lectin Modern research on lectins, carbohydrate-binding protein, started from a study on plant lectins from legume seeds. Carbohydrate-binding proteins have been found in a wide variety of mammalian organs, such as a liver, and body fluids, and are called animal lectins. Animal lectins include C-type lectin, galectin, siglec, etc. Animal lectins are characterized by a cluster effect and carbohydrate recognition known as pattern recognition. The functions of especially host defense and ER quality control mechanisms are being clarified.

β-1,3-linkage glycosyltransferase motif There is a stretch of amino acid sequence, well conserved among the enzymes, which transfer sugars by β1,3-linkage (galactosyltransferase, N-acetylglucosaminyl transferase, N-acetylgalactosaminyl transferase, etc.). Each of this enzyme family has three rows of amino acid motif and each motif spans several to dozen amino acid residues. Meanwhile, the motifs are not very well conserved in core 1 Gal-T (C1GALT1,2) or iGnT.

β-1,3-N-acetylglucosaminyl transferase This is a generic name for the enzymes transfer GlcNAc by β1,3-linkage. The reported enzymes in this group are the following: β3GnT5, which is involved in glycolipid Lc3Cer; β3Gn-T6, which synthesizes O-linked glycan core3 structure; β3Gn-T2, which functions for the synthesis of keratan sulfate; T3, which is involved in the extension of core1, T4 and T8.

β-1,4-galactosyltransferase The enzymes in this group transfer galactose by β1,4 linkage. Seven human enzymes of this group have been reported so far and they were found to contain β1,4-linkage glycosyltransferase motif in the polypeptide.

β-1,4-N-acetyl galactosaminyl transferase This enzyme transfers N-acetylgalactosaminyl residues by β1,4-linkage. The enzymes in this group are classified into three groups according to the difference in the acceptor substrate: (1) enzymes, such as β4GalNAc-T1 and T2, whose acceptor is sialylated galactose to produce GM2 or Sda antigen, (2) enzymes, such as β4GalNAc-T3 and T4, whose substrate is N-acetylglucosamine and produce LacdiNAc structure, and (3) five enzymes involved in the synthesis of chondroitin sulfate from glucronic acid as substrate.

β-1,6-*N*-acetylglucosaminyl transferase The enzymes in this group transfer *N*-acetylglucosamine by β1,6-linkage. They include enzymes, which are categorized as follows: enzymes involved in *N*-linked glycan synthesis (MGAT5 and MGAT9), enzymes that participate in the synthesis of *O*-linked glycan-core structure, such as core2 Gn-T (C2GnT1 . . . 3), IGnT, which produces I-type blood antigen, and others.

β-4 galactosyltransferase motif There is a stretch of amino acids well conserved among β-1,4 galactosyl transferase. This motif is also found in some β-1,4-linkage glycosyltransferases other than β-1,4 galactosyl transferases. The consensus sequence of this motif is WGXEDD/V/W.

β-amyloid peptide It is strongly believed that the deposition of β-amyloid peptide (Aβ) in the brain is implicated in the pathogenesis of Alzheimer disease. Aβ consists of 40–42 amino acids and is generated from β-amyloid precursor protein (APP) by the proteolytic β and γ cleavage.

β-galactose 3-*O*-sulfotransferase Sulfation of glycans at the position 3 of galactose residues has been found in glycolipids and glycoproteins in mammal, and occurs by the function of β-galactose 3-*O*-sulfotransferase (Gal3ST) family. This family constitutes four member enzymes, which were numbered according to the order of discovery. Only Gal3ST-1 (cerebroside sulfotransferase, CST) functions on glycolipid, whereas others work on glycoproteins.

β-galactoside β-galactoside is a β anomeric isomer of the galactose residue. Galectin was the lectin originally called β-galactoside-binding lectin, because it recognizes the β-galactoside structure specifically, until it was renamed Galectin in 1993.

β-galactosyltransferase β-galactosyl transferase transfers galactose residues by β-linkage. In humans, six β-1,3-galactosyltransferases, seven β-1,4-galactosyltransferases, and one galactosyl ceramide synthases are reported.

BACE1 BACE1 is a protease (β-secretase), which cleaves beta-Amyloid precursor protein (APP) at the β-site. BACE1 cleavage initiates the production of β-amyloid peptide (A β) to form amyloid plaque, which is considered to be a major cause of Alzheimer disease (AD). Therefore, BACE1 inhibitors are being developed as promising therapeutic agents for AD.

Bifidobacteria growth stimulator Milk and some other particular foods have components to stimulate the growth of bifidobacteria, which reside in the colon and have a protective role against infectious disease. Milk oligosaccharides are believed to be bifidobacteria growth stimulators. Recently, oligosaccharides having lacto-*N*-neo-biose I (e.g., lacto-*N*-neo-tetraose) have been hypothesized to be growth stimulators for *Bifidobacterium bifidus* and *Bifidobacterium longum*.

Biocombinatorial synthesis of glycan chains Biocombinatorial synthesis of glycan chains is a technique, which involves the synthesis of diverse oligosaccharide structures by combination of glycan chain primers and different kinds of cell types. By choosing appropriate glycan primers, considering cell types depending on their glycan biosynthetic pathway, this technique would be promising in synthesizing diverse kinds of glycan chains.

Bioinformatics Bioinformatics is a coinage designed in the 1990s from the combination of "biology" and "informatics". Bioinformatics and computational biology are used interchangeably. In a narrow sense, bioinformatics is concerned with biological information. In a wide sense, bioinformatics include the field of computational and statistics related to diverse biological phenomena (e.g., sequence alignment, protein structure alignment, and genome annotation).

Botulinum toxin/botulin Botulinum toxin is a neurotoxin protein produced by the bacterium *Clostridium botulinum*. Orally administered toxin is absorbed by the small intestinal epithelium and hematogenously reaches neurons, in which the toxin is endocytosized. The light chain of botulinum toxin is a protease that degrades SNAP-25, a membrane bound Q-SNARE protein. Because SNAP-25 is required for the release of neurotransmitters, such as acetylcholine, botulinum toxin suppresses the release of acetylcholine. Botulinum toxin is now used in the treatment of blepharospasm, urgency incontinence, and cosmetic surgery.

CabosDB CabosDB is an abbreviation for carbohydrate sequencing database and enables the researchers to find out information on glycan structural analysis. It includes information on oligosacharide structures and their mass analysis spectrum, lectins and their binding specificities to glycans, and glycoprotein and their glycosylated positions.

Caenorhabditis elegans (C. elegans) *Caenorhabditis elegans* is a model organism widely used in genomic and post-genomic studies. The mechanism of apotosis was elucidated from the study of *C. elegans*. RNAi and GFP technologies were developed using *C. elegans*. *C. elegans* consists of ca 1000 somatic cells and the length is about 1 mm. *C. elegans* could be cryopreserved and all the developmental processes and neural network system have been already elucidated. *C. elegans* is the only multicellular organism in which high-throughput gene disruption is possible.

Cancer metastasis Metastasis is the major factor that influences the prognosis of cancer. In cancer metastasis, cancer cells can break away from a primary tumor, enter lymphatic and blood vessels, circulate through the bloodstream, and settle down to grow within normal tissues elsewhere in the body. Metastasis can be roughly classified into hematogenous metastasis, lymphgenous metastasis, or peritoneal disseminated metastasis, depending on the metastatic path, though they are not mutually exclusive but can be related with each other, progressing simultaneously.

Capillary affinity electrophoresis Capillary affinity electrophoresis (CAE) is carried out in the electrolyte containing ligands or proteins that bind to ligands. It is a technique in which the migration patterns of interacting molecules in an electrical field are observed and used to quantify specific binding and estimate binding constant. When a carbohydrate in the mixture interacts with carbohydrate-binding protein, CAE allows determining subtle difference in affinities among the mixtures of carbohydrates. CAE also allows characterization of the structures of carbohydrates, on the basis of their affinities for an appropriate set of carbohydrate-binding proteins.

Capillary electrophoresis Capillary electrophoresis is a high-resolution separation technique that works on the basis of the different behavior of charge molecules in a high electric field (typically 10–30 kV). The technique uses a narrow (25–100 mm i.d.) capillary column that has borne silica, and its inner surface is coated with stationary phase. Thus, it provides highly efficient separations up to 1 000 000 theoretical plates. By connection with laser-induced fluorescent detector and with mass spectrometer, the technique also allows high-sensitive detection and structural characterization of samples.

Carbohydrate-deficient glycoprotein syndrome (congenital disorders of glycosylation, CDG) The standard name was changed from "carbohydrate-deficient glycoprotein syndrome" to "congenital disorders of glycosylation syndrome". In the narrow sense, CDGs are generic names of syndromes caused by genetic disorders of the enzyme related to N-glycosylation. CDGs can be distinguished into the disorder of biosynthesis of dolichol-linked oligosaccharide in the rough-surfaced endoplasmic reticulum (ER) and glycosylation of protein (CDG-I) and into that of the processing in the ER and Golgi apparatus of the N-linked glycans attached to proteins (CDG-II). The name of the disease is described by fixing alphabets on the basis of each responsible gene. The condition affects a wide variety of organs, especially the nervous system.

Carbohydrate-deficient phenomenon This is a concept proposed by Hakomori et al. as one of the patterns in the change of carbohydrates with malignant transformation. In normal cells, the structure of carbohydrates being synthesized is of the complex type. Malignant transformation causes trouble in a part of the carbohydrate synthesis, and in malignant cells, carbohydrates that have a simpler structure than in normal cells accumulate. Such a phenomenon is called carbohydrate-deficient phenomenon.

Carbohydrate library Carbohydrate library is a series of various carbohydrates or their derivatives constructed by organic synthesis, enzyme reaction, and extraction from natural resources. Besides being used as standard samples for structural analysis, carbohydrate library could be applied to a comprehensive analysis of interactions by combination with microarray technology, which would become a powerful tool for glycomics.

Carbohydrate ligands for selectins Selectin molecules mainly recognize carbohydrates as ligands, and each selectin binds to a specific ligand. E-selectin binds to sialyl Lewis x and sialyl Lewis a. P-selectin binds to sialyl Lewis x on the PSGL-1 molecule carrying sulfated tyrosine residue(s) in its N-terminal region. L-selectin binds to sialyl 6-sulfo Lewis x.

Carbohydrate microarray Carbohydrate microarray is prepared by immobilizing a variety of carbohydrates onto microarray slides such as glass plates. The microarray is typically used for the observation of the interactions between immobilized carbohydrates and fluorescent-labeled proteins. Sometimes, carbohydrate microarray is confused with carbohydrate chip. In general, the number of sugars on a carbohydrate microarray is larger than that on a sugar chip. Recently, a new type of carbohydrate microarray has been developed to observe the interaction with surface plasmon resonance (SPR) techniques, in which fluorescent labeling of protein is not necessary.

Carbohydrate receptor Carbohydrate receptor is a generic name of a receptor or a lectin binding to sugar chains in the broad sense. In enzyme replacement therapy for lysosomal disease (lysosomal storage disease), it is a receptor, such as a cation-independent mannose 6-phosphate receptor or a mannose receptor, on the target cell surface that recognizes and binds to a specific sugar chain structure and mediates internalizing the sugar chain via endocytosis and transportation into lysosome.

Carbohydrate recognition cytokine Cytokines that are small amounts of various bioactive substances bind to corresponding specific receptors and activate cells. In case cytokines bind to receptor subunits to form high-affinity complex, (cytokine)$_n$(receptor subunit)$_n$, carbohydrate recognition cytokines with lectin-like activity that trigger the recognition of receptors themselves, glycolipids close to receptors, GPI anchors, high-mannose-type sugar chains or heparan sulfate chains exist.

Carbohydrate xenoantigen Carbohydrate xenoantigens are glycan epitopes which are not found commonly in humans, but in other animals. For example, H–D antigen, Forssman antigen (GalNAcα1-3GalNAcβ1-4Galα1-4Galβ1-4Glcβ1-ceramide), Paul–Bunnell (P–B) antigen (erythrocyte glycoprotein antigen found in sheep, horse, or cow), or Galβ1-3Gal antigen are not found in humans, primates, and old world monkeys. These antigens are potential cause for immunological rejection against glycoprotein drugs prepared from non-human cells, xenotransplantation, or transplantation of human ES cells cultured under FBS (fetus bovine serum).

CD1d CD1d is an MHC class I-like molecule on antigen-presenting cells and immature thymus cells, which present the glycolipid with Galα1-linkage to T cell receptor (TCR) on NKT cells. Recently, its crystal structure was resolved and putative binding sites to glycolipids was proposed. Meanwhile, it is still disputed about which glycolipid is the in vivo ligand. CD1d is getting to be a hot topic in its relation with the activation of NKT cells.

Ceramidase Ceramidase is an enzyme, which hydrolyzes the amide bond between the sphingosine base and fatty acid in free ceramide. This enzyme is widely conserved from bacteria to humans.

It is classified into: (1) neutral ceramidase, (2) acid ceramidase, and (3) alkaline ceramidase according to the primary structure and optimum pH. The acid ceramidase exists in lysosome and the deletion of the enzyme results in Fabry's disease. The neutral ceramidase is mainly localized in the plasma membrane and modulates signal transduction through sphingolipid mediator.

Ceramide Ceramide is an *N*-acylsphingosine lipid that anchors the glycan of glycosphingolipid or phosphocholine of sphingomyelin onto the plasma membrane. Ceramides are produced by the degradation of sphingolipid or by a de novo biosynthetic pathway, and the amount of production is enhanced by the following stimuli: cytokine, heat, or ultraviolet. Ceramides are known to control a variety of signal transduction systems. Phosphorylated sphingosine is a ligand for G-protein coupled receptors (S1P, Edg).

Cerebroside sulfotransferase (CST) Cerebroside is a common name for galactosylceramide. Cerebroside sulfotransferase is a sulfatide synthase that transfers a sulfate onto the 3 position of galactose of galactosylceramide (cerebroside). This enzyme synthesizes not only sulfatide, but also lactosylceramide sulfate and seminolipid.

Chemoenzymatic synthesis To synthesize glycoconjugates, there are two major methods: chemical method and enzymatic method. Chemoenzymatic synthesis is the combination of these two methodologies, to utilize the advantages of the respective methods. For instance, the chemically synthesized glycopeptides with a monosaccharide can be utilized for synthesis of neoglycoconjugates through transglycosylation reaction of endoglycosidases.

Chondroitin sulfate Chondroitin sulfate is a glycosaminoglycan that consists of repeating disaccharide units of glucuronic acid and GalNAc. Various sulfation reactions give rise to structural diversity of this glycan. It occurs ubiquitously in most tissues, most abundantly in cartilage. Various bioactivities such as neurite outgrowth-promoting activity have been reported.

Chondroitin sulfate glucuronic acid transferase Chondroitin sulfate glucuronic acid transferase has a glucuronic acid transferase activity. The enzyme is involved in the synthesis of chondroitin (sulfate), having the repeating disaccharide unit, -3GalNAcβ1-4GlcNAcAβ1. It forms a gene family with chondroitin synthase, chondroitin polymerizing factor (ChPF) and chondroitin sulfate *N*-acetylgalactosamine transferase.

Chondroitin sulfate-*N*-acetylgalactosamine transferase Chondroitin sulfate-*N*-acetylgalactosamine transferase is involved in the synthesis of a repeating unit of chondroitin (sulfate), 3GalNAcβ1-4GlcAβ1. So far two distinct enzymes are reported, ChGn-1 and 2. It forms a gene family with chondoroitin synthetase, chondroitin-polymerizing factor, and chondroitin sulfate glucuronyltransferase.

Chondroitin sulfate synthase Chondroitin sulfate synthase is an enzyme having both *N*-acetylgalactosamine transferase and glucuronic acid transferase activities, participating in the synthesis of chondroitin (sulfate). Two kinds of enzymes, chondroitin synthase (ChSy)-1 and -2 were reported. Each ChSy can form a complex with chondroitin polymerizing factor (ChPF).

Choresteryl-α-D-glucopyranoside Choresteryl-α-D-glucopyranoside is a glycolipid where α1-3-linked glucose is attached to chresterol, often abbreviated as CGL. CGL, together with FAG (Chroresteryl-6-*O*-tetradecanoyl-α-D-glucopyranoside) or CPG (choresteryl-6-*O*-phosphatidyl-α-D-glucopyranoside), occurs in the cell wall of Helicobacter species including *H. pylori*, and is involved pivotally in growth, motility, and maintenance of morphology of these bacteria.

Chromosome mapping Chromosome mapping is to determine the position of genes on chromosomes. Fluorescently labeled DNA (e.g., FITC-DNA) is used as a probe for chromosome mapping. The position of the gene can be determined by the correlation between the chromosome band and the fluorescent signal.

C-linked glycosylation of protein Protein glycosylation was classified into two types: *N*-linked type (Asn-linked type) and *O*-linked type (Ser/Thr-linked type). Recently, tryptophan-linked mannose (C-mannosyl tryptophan) was discovered and the function is a topic of great interest.

COG complex COG is an abbreviation for conserved oligomeric Golgi complex. It is an octameric complex, which is considered to regulate the COP-I vesicular transport between Golgi stacks. Recently, this complex has received special attention after the discovery of a pathogenesis showing irregular glycosylation (CDG; congenital disorders of glycosylation) because of the deficiency in a part of the complex.

Collectin Collectin is a collective name of animal lectins bearing a conserved carbohydrate-recognition domain (CRD) as well as collagen-like domain (Gly-X-Y triplet), collagen-like lectin. It includes mannan-binding protein (MBP), conglutinin, CL-43, CL-46, surfactant protein A and D (SP-A and SP-D), CL-L1, CL-L2, and membrane type CL-P1. This protein is known to be implicated as modulators of the innate immune response, where it has a key role in the first line of defense against invading microorganisms.

Combinatorial chemistry Combinatorial chemistry is to access vast numbers of molecules efficiently in a combinatorial fashion. Combinatorial chemistry is exploited in the optimization of chemical reaction or generation of novel functional materials. Solid-phase or multi-component synthesis is a representative of combinatorial chemistry.

Complement-dependent lectin pathway The complement system is a host defense system designed to destroy pathogens. Once the complement system is activated, a chain of reactions involving proteolysis and assembly occurs, resulting in the destruction of the membranes of the pathogens. There are three categories of activation pathways: the classical pathway, alternative pathway, and lectin pathway. As compared with the classical pathway in which antibodies are required, the latter two pathways do not require antibodies. Complement-dependent lectin pathway is activated by the binding of mannan-binding protein (MBP, also called MBL) to carbohydrates on the surface of pathogens. Binding of MBP to pathogens triggers the activation of MASP (MBP-associated serine protease), which is bound to MBP, and C4, C2, and C3 subsequently undergo limited proteolysis to be activated. Ficolin, GlcNAc-binding lectin, has been recently identified as another recognition molecule of the lectin pathway.

Complex-type N-glycan Vertebrate N-glycans are classified into three types: high-mannose, hybrid, and complex. In case of the complex-type N-glycan, both α3- and α6-linked mannose residues are substituted with GlcNAc residues. Most secreted and cell surface glycoproteins have complex-type N-glycan.

Conditional knockout mice A knockout mouse is a genetically engineered mouse that has a particular gene inactivated by replacing or disrupting it with an artificial piece of DNA. Analysis of knockout mice will provide researchers with valuable clues about the gene function in vivo. *Conditional knockout strategy* works on the basis of a tissue-specific inactivation of the gene of interest, and it is a powerful technique especially when constitutive knockout mice are lethal. This can be achieved by means of Cre-Lox recombination system, which involves the targeting of a specific sequence of DNA having lox P sites (consisting of 34 bp) and splicing it with Cre protein, a site-specific DNA recombinase.

Congenital muscular dystrophy The term congenital muscular dystrophy is used to describe the muscular dystrophy present at birth. Muscular dystrophies are characterized by progressive skeletal weakness, defects in muscle proteins, and the death of muscle cells and tissues. Fukuyama congenital muscular dystrophy is a form of muscular dystrophy mainly described in Japan.

Core-3 glycans Core-3 glycans is one of the core structures of mucin type, O-linked glycans, GlcNAcβ1-3GalNAcα1-Ser/Thr. It is found in normal stomach and intestine, whereas the level of core 3 structure is reduced in cancer tissues.

Dendrimer Dendrimers are repeatedly branched molecules showing dendric architectures. Dendrimers are usually sphere or sector structures. The number of branching is called "generation". Dendrimers are composed of three elements: core, interior, and surface and have the potential to express a variety of functions by designing each element. With great attention, dendrimers are regarded as a new nanomaterial in glycotechnology and other material science fields.

Diabetes mellitus Diabetes mellitus is a metabolic syndrome characterized by chronic high blood sugar (hyperglycemia), which is related to genetic factor and environmental factors such as lifestyle. Diabetes mellitus is characterized by a deficiency in the effect of insulin, which is caused by absolute or relative insufficiency of insulin and/or decrements in sensitivity to insulin. Diabetes mellitus is roughly classified into type 1, caused by auto-immune mechanism, and type 2. It is reported that, in carbohydrates, the activation of the hexosamine pathway and errors in ganglioside metabolism are the factors of insulin resistance.

DNA microarray DNA microarray is a small chip made of glass or some other chemical materials on which various DNA fragments are spotted and arranged, and is used to analyze the expression of multiple genes simultaneously. Thus, it can tell the gene expression profile. Generally, it is used to compare the relative expression levels of those genes in a couple of samples by comparing the signal intensity after the hybridization of cDNA from both samples labeled with different fluorescent dyes.

Drosophila *Drosophila (Drosophila metanogaster)* is a small fly that is 2–4 mm long. Full-length genome of *Drosophila metanogaster* has been sequenced, and *Drosophila* is extensively used as a model organism in many scientific fields. Recently a lot of glycogenes have been identified and the functional analyses are now in progress.

Dystroglycan Dystroglycan is a component of dystrophin-glycoprotein complex formed in sleletal muscle. Dystroglycan is encoded by a single gene and the protein is cleaved into α and β subunits by posttranslational processing. α-Dystroglycan is an extracellular glycoprotein and binds to laminin. On the other side, β-Dystroglycan is a transmembrane glycoprotein that binds to dystrophin in the cytosol. Lack of dystroglycan in knockout mice results in embryonic lethality. Lack of dystroglycan in conditional knockout mice, specifically in the brain, peripheral nerve, or skeletal muscle, avoids embryonic lethality, resulting in abnormalities of each tissue.

EGF domain EGF is an abbreviation for epidermal growth factor. EGF domain is an amid acid motif homologous to epidermal growth factor (EGF) and recently it was shown that sometimes this domain is O-fucosylated and/or O-glucosylated.

EGF receptor EGF receptor is a type I transmembrane protein composed of 1186 amino acids and has 11 possible N-linked glycosylation sites. It transduces signals to the downstream molecules by dimerization and autophosphorylation on stimulation by the ligand (e.g., EGF) binding. PhospholipaseCγ/CaMK/PKC pathway and Ras/Erk cascade are some of the well-known downstream pathways of the EGF receptor.

Endo-α-N-acetylgalactosaminidase It is an enzyme that cleaves mucin-type glycans from glycoproteins. First found in the culture media of *Diplococcus pneumonia* in 1976, similar activity was observed in the culture media of *Clostridium* or *Alcaligens*. These enzymes have very strict substrate specificity, and only act on O-glycans with disaccharides, Galβ1-3GalNAc. On the other hand, new type of endo-α-N-acetylgalactosaminidase with broader glycan specificity has recently been isolated from *Streptomyces*, which has more potential to become a powerful tool reagent.

Endo-β-N-acetylglucosaminidase (ENGase) ENGase is a glycosidase that cleaves a glycoside bond in the proximal N,N'-diacetylchitobiose (GlcNAcβ1-4GlcNAc), releasing N-glycans from glycopeptides/glycoproteins. Various distinct enzymes with different glycan specificity have been identified. For instance, an enzyme called Endo H predominantly acts on high mannose glycans, whereas another enzyme called Endo D can also work on some complex-type glycans. On the basis of primary structures there are two types of ENGase: one is categorized as glycoside hydrolase family 18 (e.g., Endo H, Endo F), whereas the other is categorized as glycoside hydrolase family 85 (e.g., Endo D, Endo M and human cytosolic ENGase).

Endo-β-*N*-acetylglucosaminidase M (Endo M) ENGase isolated from *Mucor hiemalis*, Endo M, has relatively broad substrate specificity and can release *N*-glycans from all types of glycans (high-mannose; hybrid and complex type). This enzyme also has high transglycosylation activity, and is used as a tool reagent for chemoenzymatic synthesis of neoglycoconjugates.

Endocytosis There are two major pathways in eukaryotic cells for material transport. One of the pathways is called endocytosis, a process whereby cells absorb materials from the outside of cells by engulfing them with their cell membrane. Mainly the endocytosed materials are transported to lysosome via endosome. On the other hand, exocytosis is a process by which cells direct materials (intracellular vesicles) out of the cell membrane.

Endoglycoceramidase Endoglycoceramidase, also called ceramide glycanase, is an endoglycosidase, which cleaves glycoside bonds between glycans and ceramide of glycosphingolipids, thereby forming oligosaccharides and ceramides. It was originally identified in *Rhodococcus* sp., and later in leech, short-necked clam, and jellyfish. This enzyme can also catalyze transglycosylation reactions.

Endoglycosidase An endoglycosidase is an enzyme that releases oligosaccharides from glycoconjugates (glycoproteins, glycolipids, and proteoglycans). In a broader definition, it also includes enzymes that act on polysaccharides, such as chitinase or glucanase. These enzymes are used as tool reagents to study the structures and functions of glycans, because they can release glycans from aglycons under milder conditions compared with chemical methods. The transglycosylation activity of these enzymes is also utilized to form neoglycoconjugates.

Endosialidase Endosialidase specifically recognizes α2,8-linked oligo/polysialic acids, cleaving the glycoside bond between sialic acids. A soluble enzyme derived from bacteriophage K1F (Endo-*N*) catalyzes the following reaction: $(\rightarrow Neu5Acyl\alpha2\rightarrow)n\text{-}X$ $(n > 5)\rightarrow(\rightarrow Neu5Acyl\alpha2\rightarrow)_{2\text{-}4} + (\rightarrow Neu5Acyl\alpha2\rightarrow)_2\text{-}X$. Endo-*N* is a useful reagent to confirm the occurrence of oligo/polysialic acids, and also to analyze their function.

Endosulfatase Endosulfatase is an enzyme that acts on polysaccharides, releasing sulfate. The best known enzyme is the one acting on 6-*O*-sulfates of heparan sulfate. This enzyme has been extensively studied in terms of the relationship with development and cancer formation, because the action of this enzyme results in upregulation of Wnt signaling concomitant with inhibition of FGF signaling pathway.

Enzyme cue synthesis strategy This is a strategy for the simultaneous production of glycans with various structures in a single tube by quitting sequel glycosylating reactions in the middle of each step. First, the glycan synthesis is aborted in the middle to have a mixture of products and unreacted materials. Then, the next glycosylation step is again aborted before completion to have a mixture of the four variants of the glycosylated materials, e.g., first products and unglycosylated materials in the first reaction and their reacted products.

Enzyme replacement therapy Absence or deficiency of enzyme results in accumulation of excess substrates or toxic metabolic intermediates. Enzyme replacement therapy is a medical treatment replacing an enzyme in patients in whom the particular enzyme is absent or deficient. Clinical application of enzyme replacement therapy is performed for lysosomal enzyme disorder and adenosine deaminase deficiency (a kind of severe combined immunodeficiency).

ERAD ERAD is an abbreviation for endoplasmic reticulum-associated (protein) degradation. If proteins newly synthesized in the ER cannot fold properly, such misfolded proteins are retrotranslocated from the ER into the cytosol, and finally they will be degraded by proteasome in the cytosol. This protein degradation cascade is called ERAD and is conserved from yeast to human.

ERQC ERQC is an abbreviation for endoplasmic reticulum quality control. Newly synthesized proteins are allowed to proceed to the vesicular transport system when they acquire proper folding, whereas misfolded and/or misassembled proteins are retained at the ER. ERAD pathway of misfolded proteins is also included in ERQC.

Evolution Diversity of oligosaccharide structures among various biological species is acquired through evolutionary process. The evolution of sugar chains was not well investigated because of the lack of structural data on oligosaccharides. However, a lot of glycosyltransferase genes have been cloned and the evolution of sugar chains is now well studied with the molecular evolution approach.

Expression cloning Expression cloning is a technique in DNA cloning that uses expression vectors to generate a library of clones, with each clone expressing one protein. This expression library is then screened for the property of interest, according to the antigen epitope or enzyme activitiy of the expressed proteins.

EXT gene family EXT gene family is composed of five genes: the EXT1 and EXT2, which are responsible for hereditary multiple exostoses (HME), and their homolog EXT-like genes (EXTL1, EXTL2 and EXTL3). It is not known if there is any relationship between HME and EXTL1 . . . L3.

All these five EXT genes encode glycosyltransferases for the synthesis of heparan sulfate.

Fluorescent labeling of glycans Reducing end of glycan chains can be labeled by organic compounds, forming chemically stable fluorescent derivatives. This derivative achieves highly sensitive detection through HPLC or mass spectrometry. Especially, this method is often used for structure analysis of glycans using reversed-phase HPLC.

Fluorescent polarization measurement Fluorescence polarization, excited with polarized light and emitted from the solution of the fluorescent-labeled molecules, is depolarized by the Brownian motion of the molecule itself. Because the Brownian motion can be correlated with the size of the molecule, measurement of the real-time interaction of fluorescent probes with other macromolecule is possible through the fluorescent polarization measurement. This method achieves real-time measurement of biomolecular interaction analysis without, not like other chip-based methods, a problem of non-specific interaction. See also "Single molecule fluorescent measurement (fluorescent correlation spectroscopy)".

Frontal affinity chromatography Frontal affinity chromatography (FAC) is a quantitative affinity chromatography, developed in 1974 by Kasai et al., which can accurately determine affinity constants between biomolecules such as enzymes–substrate analogues and lectin–oligosaccharides. FAC is more advantageous than other methods for the analysis of relatively weak interactions, such as sugar–protein interaction.

GALAXY (GlycoAnaLysis by the three AXes of MS and chromatographY) GALAXY is a web application for supporting structural analysis of N-linked glycans. Its database includes 2-D/3-D HPLC maps and molecular weight data for about 500 pyridylaminated oligosaccharides on the Intenet. The URL is http://www.glycoanalysis. info/

Galectin Galectins are defined as lectins having both β-galactoside binding and amino acid sequence similarity. It occurs not only in animals, but also in fungi (mushroom). This protein is known to be involved in quite diverse biological events, i.e., development, differentiation, apoptosis, growth, inflammation, and cancer metastasis.

Ganglioside Gangliosides are defined as sialic acid-containing glycoshingolipids. Most of the gangliosides are composed of ganglio-series, whereas it also includes sialic acid containing lacto-, neolacto- and globo series glycosphingolipids. It is most abundantly found in neuronal tissues, and it has been shown that gangliosides are involved in various biological events such as development and differentiation.

Gb3/CD77 Gb3 stands for globotriaosylceramide (Galα1,4Galβ1,4Glc-ceramide). It is often described as Gb3/CD77, and it was also named as CD77 antigen after being found on Burkitt lymphoma and immature B lymphocyte.

It is also known as a receptor for verotoxin, which is produced by pathogenic *Escherichia coli* O157. In humans, expressions are observed on the erythrocytes, kidney cells, splenocytes, and vascular endothelial cells. Its glycan structure is produced by α1,4 galactosyl transferase.

GD2 GD2 stands for GalNAcβ1,4 (NeuAcα2,8NeuAcα2,3) Galβ1,4Glc-ceramide. It is considered to be a cancer-related antigen, because it is often found in embryonic brain and also highly expressed in tumor derived from neuroectoderm, such as melanoma, neurblastoma or glycoma. It was also reported to be expressed in a subset of T cell lymphoma or small cell lung carcinoma.

The glycan moiety is synthesized by the action of β-1,4-N-acetylgalactosaminyltransferase.

GD3 GD3 stands for NeuAcα2,8NeuAsα2,3Galβ1,4Glc-ceramide. It is highly expressed on embryonic brain, melanoma, T cell lymphoma or activated T cells. This molecule is proposed to be the target for the antibody therapy for melanoma. Recently it was reported that this antigen causes the expression of cancerous phenotypes and also induces apoptosis.

This glycan structure is produced by ST8SiaI.

GDP-L Fucose synthase GDP-Fucose, which is a donor substrate of fucosyltransferase, is synthesized in cytosol and translocated into Golgi by the function of GDP-fucose transporter. There are a couple of GDP-Fuc synthesis pathways known: one is "salvage pathway", which includes Fuc kinase and GDP-Fuc pyrophosphorylase, and another is "de novo pathway", which includes GMD and FX.

Gene expression profiling Gene expression profiling is a whole picture of expression pattern of multiple genes. This can be obtained by methods such as SAGE (serial analysis of gene expression) or DNA microarray technique. Normally, these data are processed by clustering analysis, allowing us to envisage the character of cells and/or function of genes of interest. This method is also utilized to identify genes responsible for differentiation or malignant transformation of cells by comparing the expression profile between different cells.

Gene knockdown Gene knockdown is a method of downregulation of gene expression. In mammalian cells, short interfering RNA (siRNA) or short hairpin RNA (shRNA) are often used for this purpose, because such RNAi-based method will avoid interferon response.

Gene knockout Gene knockout is a method of complete inactivation of gene function. In the case of mouse, KO animal is generated through replacement of the target gene to, in most cases, a marker gene, which confers resistance to a certain toxic agent for selection. Tissue-specific KO mice can also be generated using Cre/loxP system.

Genetic polymorphisms Genetic polymorphisms arise from individual difference in DNA (genomic) sequence. In blood typing (ABO system, P system or Ii system), variable glycan antigens are formed because of single polymorphisms, resulting in a change in substrate specificity or reduced/loss of activity of responsible enzymes.

Glucosyl cereamide Glucosyl ceramide is a glycolipid in which a glucose residue is attached to ceramide. This glycolipid distributes ubiquitously in cells and serves as precursors of various glycolipids. It can also regulate the level of free ceramide in cells, controlling cellular signals. This lipid is synthesized on the cytosolic face of the endoplasmic reticulum by glucosyl ceramide synthetase.

Glycan database Recently, a variety of glycan-related databases have been constructed: (1) Database about glycan structures, glycan-related enzymes such as glycosyltransferases, anti-glycan antibodies, and lectins. (2) Web applications that support the analyses of two-dimensional HPLC mapping and mass spectrometry. (3) Integrated database of these databases.

Glycoamidase A Discovered in almond seeds in 1977, this enzyme is the first glycoamidase (peptide : N-glycanse) described in nature, having an optimal pH of 4. See also "peptide : N-glycanase".

Glycocluster effect The term, glycocluster effect, is used to represent the sugar-recognition mode of animal lectins. Although animal lectin has weak affinity for monosaccharide, apparent affinity of the lectins for monosaccharide-coated protein (neoglycoprotein) becomes high, depending on the number of the monosaccharides attached to a carrier protein. Animal lectins show high affinity toward the assembled non-reducing sugar residues.

Glycoconjugates Glycoconjugates represent carbohydrates, which are covalently attached to a nonsugar moiety (lipids or proteins). The major glycoconjugates are glycoproteins, glycolipids, and proteoglycans.

Glycoepitope (Glycotope) An epitope is the part of an antigen that is specifically recognized by antibodies. Generally, oligosaccharide-binding antibodies recognize several monosaccharides as an epitope (glycoepitope). Anti-carbohydrate antibodies can be produced because of the high antigenicity (immunogenicity) of the glycans. A variety of anti-carbohydrate monoclonal antibodies are now produced.

GlycoEpitope DB There have been tremendous numbers of mono- or polyclonal antibodies produced and used for glycan function analyses and medical examinations by tissue staining, immunoprecipitation, Western blotting, or ELISA.

Thus GlycoEpitope DB was innovated to store the Glycoepitopes or glycan structure-specific antibodies in an organized manner. Its Internet URL is http://www.glyco.is. ritsumei.ac.jp/epitope/

Glycogene Glycogene is a word that was coined by Narimatsu (AIST, Japan) several years ago. It is a family of genes that codes synthesizing, degrading, or binding of proteins to sugar chains. Glycogene includes genes associated with (1) glycosyltransferases, (2) sulfotransferases, which transfer a sulfate group to a sugar chain, (3) sugar–nucleotide transporters, (4) sugar–nucleotide syntheses, etc.

GlycoGene DB (GGDB) GGDB is an Internet database for the genes involved in glycosylation, such as glycosyltransferases, sulfotransferases or nucleotide-sugar transporters.

Glycolipid Glycolipids are a series of glycoconjugates synthesized by attaching glucose and/or galactose to a ceramid. The structures from the former are predominant. Sialic acid, GlcNAc, or Gal attach to a lactoceramide, and diverge into ganglio, lacto/neolacto, or globo series, respectively. They are mainly expressed on the cell surface and are thought to be especially involved in the tissue or the cell. Besides these, glyceroglycolipids also exist.

Glycomics The word "glycome" is a relatively new word, first defined at the end of the twentieth century as the entire complement of sugars, the third biological chains next to DNA and proteins, of an organism, just like genomes for DNA and proteomes for proteins. The research that studies glycomes is referred to as glycomics.

Glycopeptide A glycopeptide is a complex molecule consisting of carbohydrate and peptide, and is one of the glycoconjugates. A glycopeptide is derived from the digestion of glycoprotein by a protease.

Glycoprotein Glycoproteins, complex morecules of carbohydrates and proteins, are one of the glycoconjugates. Proteoglycans are often classified into different categories.

Glycosaminoglycan Glycosaminoglycans (GAG), previously known as mucopolysaccharides, are long unbranched polysaccharides consisting of repeating disaccharide units. Examples of GAGs include heparin/heparan sulfate, chondroitin sulfate/dermatan sulfate, keratan sulfate, and hyaluronan. Most of the GAGs, except for hyaluronan, are sulfated and are covalently linked to a protein to form proteoglycans.

Glycoside hydrase Glycoside hydrase is a generic term for enzymes involved in the hydrolysis of oligosaccharides and exists in lysosomes and cytosol. Lysosomes contain various glycoside hydrases and the gene deletion of glycoside hydrase results in lysosomal deficiency caused by accumulation of glycoconjugates.

Glycosphingolipid Glycosphingolipid is a generic name for glycolipid-containing sphingosine base. Glycosphingolipids usually form microdomains at the outer plasma membrane in a cholesterol-dependent or -independent manner. Glycosphingolipid-enriched mirodomain can regulate the functions of many membrane proteins, e.g., signal transduction across the membrane or cell–cell adhesion. Glycosphingolipid can function as a receptor for pathogenic bacteria, virus, and their toxins. Glycosphingolipid containing sialic acid is called ganglioside.

Glycosyl fluoride Glycosyl fluoride is an oligosaccharide replaced with fluoride. Glycosyl fluoride derivatives, in which C-1 hydroxyl group is substituted with fluoride, are useful synthetic intermediates for organic chemical- and enzymatic glycosylation.

Glycosyltransferase Glycosyltransferases are an enzyme family related to the biosynthesis of sugar chains located in the Golgi apparatus. There are approximately 300 glycosyltransferase genes, and two-thirds of them have been cloned to date. Glycosyltransferases exist in only very small amounts in the body and have specific expression patterns in a variety of organs. Moreover, the expression changes during development, differentiation, malignant transformation, etc.

Glycovirology Glycovirology is a research field covering both glycobiology and virology. The First International Meeting on Glycovirology was held in Sweden in June 2007. Sugar chain will surely be a very important target for research and drug discovery in the twenty-first century because of its close relationship to viral infection. Development in the field of glycovirology and fostering specialists in this research are derived.

GM2 GM2 stands for $GalNAc\beta1,4(NeuAc\alpha2,3)$ $Gal\beta1,4Glc$-ceramide. It is expressed in embryonic brains, melanoma, glioma, and lung carcinoma; it is often observed to express epithelial carcinoma cells. This structure is synthesized by β-1,4-N-acetylgalactosaminyltransferase.

GM2/GD2 synthase This enzyme catalyzes the reaction of the synthesis from GM3, GD3 into GM2, GD2. It is also called β-1,4-N-acetylgalactosaminyltransferase (β4GalNAc-TI). It also can catalyze the GA2 synthesis reaction from lactosylceramide. The gene encoding this enzyme was cloned in 1992 for the first time among the glycolipid synthase. This enzyme functions on the synthesis of every single complex-type ganglioside.

GM3 GM3 stands for $NeuAc\alpha2,3Gal\beta1,4Glc$-ceramide. It is located at the beginning of the synthesis pathways of sialylated glycolipids, and is synthesized by ST3SiaV. It is not highly expressed in the neuronal cells even with the high levels of its synthesis, because the product will be transformed into other glycolipids. It is observed in any of normal cancerous cells. Recently, its significance for the regulation of signaling pathways of EGF or insulin has been reported.

Golgi apparatus Golgi apparatus (also called Golgi body or Golgi complex) is an organelle composed of membrane-bound stacks called cisternae. It is found in most eukaryotic cells, bearing various glycosyl transferases and related proteins (e.g., sugar nucleotide transporters, etc.), and thus is involved importantly in glycosylation.

Growth factor Growth factor is a protein that promotes cellular proliferation and differentiation. Fibroblast growth factor (FGF) is one of the well-known growth factors.

The interaction with heparan sulfate is usually essential for the functional expression of the growth factor.

HCDM (human cell differentiation molecules) Originally known as HLDA (human leukocyte differentiation antigen), they are generally called the CD (cluster of differentiation) molecules. HLDA was previously restricted to being the cell surface molecules on leukocytes and its precursors, but currently the HCDM represents any differentiation antigens on other cell types, such as platelets, erythrocytes, endothelial cells, etc., and even includes the intracellular markers. Therefore, with the ever-changing field of studies surrounding HLDA, it was decided to change the name to HCDM.

H–D antigens Hanganatziu–Deicher antigen is named after the researchers who found the antigen. It has been known that patients who are treated with horse, sheep or goat serum often experience anaphylactic shock on the second treatment. The antigen responsible for this was found to be NeuGc-containing glycans, which is normally absent in human tissues.

Helicobacter pylori *Helicobacter pylori* is a Gram-negative, helical shaped bacteria having several flagella lives in the stomach. Urease produced from *H. pylori* metabolizes urea to carbon dioxide and ammonia (which neutralizes gastric acid). *H. pylori* infection is strongly related to not only chronic gastritis and peptic ulcer, but also stomach cancer and gastric lymphoma. Finding of *H. pylori* shed light on the Gastroenterology and infectious disease society. In 2005, Dr. Marschall and Dr. Warren, who found *H. pylori* were awarded the Novel prize in Medicine.

Heparan sulfate (HS) Heparan sulfate (HS), a member of glycosaminoglycan, consists of repeated disaccharide units of glucuronic acid (GlcA) and GlcNAc. HS is found on the cell surface and extracellular matrix of most animal cells. HS modification, in which the *N*-acetyl group of GlcNAc and various hydroxyl groups are partially sulfated, create structural heterogeneity and thereby enables the HS interaction with a variety of biologically active proteins such as growth factors, morphogenic ligands, and cyokine.

Heparan sulfate *O*-sulfoglycosyltransferase After the heparan sulfate (HS) chain is modified by *N*-deacetylase/*N*-sulfotransferase (NDST) and glucronyl C5 epimerase, which converts GlcA to iduronic acid (IdoA), HS *O*-sulfation occurs at various positions. Heparan sulfate *O*-sulfotransferases are cateogorized into three types: HS2ST(2-*O*-sulfation of GlcA/IdoA), HS6ST (6-*O*-sulfation of GlcN), and HS3ST(3-*O*-sulfation of GlcN).

Heparin Heparin is a highly sulfated glycosaminoglycan (GAG), which is composed of a 2-*O*-sulfated iduronic acid, and 6-*O*- and *N*-sulfated glucosamine, IdoA(2*S*)-GlcNS(6*S*). Heparin has strong anticoagulant activity, and thereby is used for anticoagulation in surgery. Heparin's exact physiological role is still unclear, because blood anti-coagulation is achieved mostly by endothelial cell-derived heparan sulfate proteoglycans. It has been proposed that, rather than anticoagulation, the main purpose of heparin is a defensive mechanism at sites of tissue injury against invading bacteria and other foreign materials.

Heparin-binding growth factor Heparin-binding growth factor represents a growth factor family having heparin-binding activity and contains basic FGF, HB-EGF, and midkine. Heparan sulfate functions as a co-receptor for the high affinity receptors of various growth factors.

Hidrazinolysis In a narrow sense, hidrazinolysis is a technique that releases glycan chains. Free N-linked glycan chains can be released by heating glycoproteins in hidorazin. Free amino groups caused by hidrazinolysis are usually acetylated. There is a report of a modified hidrazinolysis technique to obtain free O-linked glycan chains.

High endothelial cell High endothelial cell is expressed in high endothelial venues found in secondary lymphoid organs such as lymph nodes. High endothelial cells are characterized by their cuboudal form as opposed to squamous cells found in regular endothelial venues. Lymphocytes expressing L-selectin can move between high endothelial cells into the secondary lymphoid organs. L-selectin molecules weakly interact with specific oligosaccharides expressed on high endothelial cells.

High-mannose type glycan High-mannose type glycan is a kind of N-linked glycans that exist in eukaryotes. The glycan normally consists of 5–9 mannose and two N-acetylglucosamine residues. It is a biosynthetic precursor of hybrid-type and complex-type N-glycan. The high-mannose glycan seems to be related to the intercellular transport of glycoproteins.

HNK-1 antigen HNK-1 antigen is a carbohydrate antigen recognized by anti HNK-1 (human natural killer-1) antibody. HNK-1 antigen, which contains 3-O-sulfated glucronic acid residue at its non-reducing terminus, is mainly expressed in neuronal cells.

HNK-1 Glucronyltransferase This enzyme transfers glucuronic acid to a terminal galactose residue of N-acetyllactosamine structure in glycans, forming HNK-1 epitope structure.

HPLC (high-performance liquid chromatography) HPLC is a form of column chromatography used frequently in biochemistry and analytical chemistry. It is also referred to as high-pressure liquid chromatography. The use of small-particle resins makes it possible to be compatible for high pressure, giving the components less time to diffuse within the column and thus leading to improved resolution.

Hyaluronidase Hyaluronidases are a family of enzymes that degrade hyaluronic acid by cleaving $\beta1,4$-glycosidic linkage between GlcNAc and GlcUA. Bacteria hyaluronidase catalyzes the elimination reaction to cleave hyaluronic acid, and animal-type hyaluroniade catalyzes the hydrolysis reaction. Both types of enzymes have activities to cleave chondroitin sulfuric acids. In humans, seven hyaluronidase-like genes are identified, and some of the gene products may be involved in cancer progression.

Hyaluronic acid/hyaluronan/hyaluronate Hyaluronic acid is a non-sulfated glycosaminoglycan that has a repeated linear structure of disaccharide unit of (-4GlcUA$\beta1,3$GlcNAc$\beta1$-). Hyaluronic acid is synthesized by hyaluronan synthases. Hyaluronic acid generally consists of several thousand disaccharide repeats in length, and range in size from 10^3 to 10^6 Da depending on the distribution or the physiological condition. Hyaluronic acid is also one of the major components of the extracellular matrix, and contributes to tissue formation and regulation of diverse cellular functions.

Hybrid-type glycan Hybrid-type glycan is a type of *N*-linked glycans that exists in multicellular eukaryotes. The hybrid-type glycan contains common structural features of high-mannose type and complex-type glycans.

Hypoxia-inducible factor (HIF) Hypoxia-inducible factors (HIF) are transcription factors that respond to decrease in available oxygen in the cellular environment. HIFs have the ability to induce a series of genes to adapt to or resist hypoxia. In the normal oxygen condition, the α subunit of HIF is localized in the cytosol, and the degradation of proteasome occurs after rapid ubiquitination of the protein. In low-oxygen conditions, the degradation of the α subunit is stopped and the protein binds to the β subunit after nuclear translocation. The αβ complex induces the transcription of a series of genes with hypoxia-responsive element.

IgA nephropathy IgA nephropathy is primarily characterized by deposition of IgA in glomerulus, and the most frequently seen symptoms are proteinuria and hematuria. Berger and Hinglais, in 1968, were the first to describe IgA deposition in this form of glomerulonephritis. It is commonly seen in Japan, and as much as 40% of the patients end up with the introduction of dialysis because of chronic, progressive renal failure. It is estimated that 30% of patients newly introduced to dialysis treatment have this disease. Though the direct cause of this disease is still unclear, the involvement of IgA-containing protein complex or abnormality of glycans on IgA is predicted.

IgA1-binding protein IgA1-binding protein (IgA1-BP) represents serum proteins bound to asialo- and agalacto-IgA1. These proteins can be separated from serum through asialo-, agalacto-IgA1 affinity chromatography. As the IgA1 content in IgA1-BP is significantly higher in IgA nephropathy patients, the relationship between IgA nephropathy and glomerular deposited IgA is predicted.

Ii blood type antigens Allo antigen determined by the glycan structures on erythrocytes. The polylactosamine structure [(Galβ1-4GlcNAcβ1-3)n] is called i antigen, whereas the branched structure of GlcNAcβ1-3(GlcNAcβ1-6)Galβ1-4 is called I antigen. These antigens are expressed as blood type antigens on erythrocytes as well as other tissues.

Immunostimulating glycoconjugate Innate immunity recognizes the specific substances of microorganisms (pathogen-associated molecular pattern: PAMP) and activates the immune system. Most of the bacterial glycoconjugates are recognized as PAMP and possess immunostimulating activity. Although most of these glycoconjugates are high molecular weight compounds, a minimum unit for the activity is usually a part of them. Peptidoglycan, lipopolysaccharide of Gram-negative bacteria and lipoteichoic acid etc., are typical of *immunostimulating glycoconjugates*.

Influenza virus It is an RNA virus containing eight pieces of segmented negative-sense single-strand RNA coated with a lipid bilayer structure called envelope. There are three types of influenza virus: influenza virus A, B, and C, on the basis of the antigenicity of the proteins. Influenza virus A can be isolated from not only humans, but also from animals such as domestic fowls, aquatic birds, pigs, or horses. Influenza virus A is thus regarded as zoonosis, and causes an outbreak or gives rise to human influenza pandemics. Influenza virus A and B contain hemaglutinin (H), which recognizes sialoglycoconjugates, and neuraminidase (N) spike. In Influenza virus A, there are 16 subtypes for H (H1–16) and 9 subtypes of N (N1–9) so far known.

KEGG DB KEGG (Kyoto Encyclopedia of Genes and Genomes) is a database of biological systems, consisting of genetic building blocks of genes and proteins, chemical building blocks of both endogenous and exogenous substances, molecular wiring diagrams of interaction and reaction networks, and hierarchies and relationships of various biological objects.

Keratan sulfate Keratan sulfate is any of several glycosaminoglycans that have been found especially in the cornea, cartilage, and brain. Keratan sulfate is a glycan polymer that consists of repeating disaccharides, [Galβ1-4GlcNAc(6-O-SO$_3$H) β1-3], attaching to the core proteins via either N-linked or O-linked glycans. In the cornea, it has been shown that this glycan provides water retention capability.

Laboratory automation Laboratory automation represents automation of all the processes performed in a laboratory such as synthesis, purification, and analysis of diverse compounds, to increase productivity, reduce lab process cycle times, and elevate experimental data quality. The most widely known technology is laboratory robotics. More recently, the field of laboratory automation comprises many different automated laboratory instruments, devices, softwares, and algorithms.

Lectin Lectin is any of group of carbohydrate-binding proteins or glycoproteins except for enzymes, antibodies, and transporter, which specifically bind to carbohydrates. Orginally, lectin was restricted to multivalent proteins capable of agglutination. However, today lectin is used in a broad sense to denote all types of carbohydrate-binding proteins that do not catalyze reactions with their ligands.

Lectin affinity Affinity for lectin is often used to purify glycoproteins or determine the interactions with carbohydrates that have complex structures Today, lectin affinity is also applied to profile characteristic structures of carbohydrates.

Lectin database Web-database, one of the sub-database in CabosDB (Carbohydrate sequencing database), consists of information on the features of more than 200 lectins and on the binding constants of more than 100 standard carbohydrates to them. On the basis of the database, useful lectins for profiling carbohydrate structures have been retrieved and the interactions between carbohydrates and lectins have been examined from various directions.

Lewis blood group antigen Lewis blood group antigens are the molecules expressed on the surface of human red blood cells. However, they are not produced by the erythrocyte itself. Instead, they are components of body secretions, and are subsequently adsorbed onto the surface of erythrocytes. They are also expressed on many other tissues including digestive tract and tumor cells. There are two types of Lewis antigens, Lewis a and Lewis b.

Liquid chromatography-mass spectrometry (LC/MS) It is an analytical system to analyze compounds separated by liquid chromatography directly subjected to electrospray ionization-mass spectrometry.

Lymphocytes homing Lymphocytes homing indicate that native lymphocytes carried to the blood migrate into specific microenviroments within secondary lymphoid tissues (lympho node, Peter's patches, and spleen). Then the lymphocytes recirculate through these sites until they either die or encounter their specific antigen. Memory and effector T cells can efficiently extravasate in extralymphoid inflammatory sites, with subsets displaying targeted trafficking through, for example, inflamed skin, and intestinal mucosa. Therefore, this tissue-selective lymphocyte homing is also regarded as lymphocyte homing.

Lysosomal enzymes Lysosomal enzymes are hydrolases responsible for breaking down complex chemicals (biological substances or chemicals derived from microbes) in an organelle called lysosome.

Mannan-binding protein (MBP) Mannose or mannan-binding protein (MBP), also called mannose-binding lectin, belongs to the collectin family in the C-type lectin. MBP binds to terminal non-reducing sugar residues, mannose, N-acetylglucosamine, or fucose, in a calcium-dependent manner. MBP is produced by the liver and secreted into circulation. It is involved in innate immunity, and its binding to microorganisms results in activation of the complement system. MBP has an oligomeric structure, built of subunits that contain identical peptide chains of 32 kDa each.

Mass spectrometry Mass spectometry is an analytical technique used to obtain structural information on the target molecules. The main steps in measuring are (1) ionization of the sample, (2) separation of ions with different masses (m/z), and (3) detection of the number of ions in each mass.

Matrix-assisted laser desorption/ionization (MALDI) MALDI is a soft ionization technique used in mass spectrometry. Soft ionization technique affords molecular ions of biomolecules such as proteins, peptides, and carbohydrates, which tend to be fragile and quickly lose structure when ionized by conventional ionization methods. A matrix is used to protect the molecule from being destroyed by the direct laser beam. In the procedure, the sample solution is mixed with a large excess matrix solution, and the aliquots of the resulting mixture are spotted on a target plate for crystallization. Finally, the laser is fired at the crystals for ionization.

Methylation analysis Methylation analysis is a chemical approach for determining linkage position of the monosaccharide residues in an oligosaccharide. This method works on the basis of the acid stability of methyl ethers and the acid lability of glycosidic linkage. First, a stable methyl group is introduced on each free hydroxyl group of the oligosaccharide. The glycosidic linkages are then cleaved, producing individual monosaccharide residues with new free hydroxyl groups that appear at the positions that were previously involved in a linkage. The monosaccharides are reduced after N-acetylation to produce volatile compounds and are analyzed by gas–liquid chromatography coupled to a mass spectrometer.

MGL/CD301 MGL/CD301 is the macrophage galactose-type, C-type lectin. In humans, a single gene is located in chromosome 17, whereas mouse has two orthologous genes in chromosome 11. This type II transmembrane molecule is expressed in the bone marrow-derived macrophages or immature dendritic cells. C-terminal carbohydrate recognition domain binds to Gal and GalNAc as monosaccharides. This protein is involved in carbohydrate-dependent endocytosis of antigens and/or cellular trafficking. MGL-KO mice exhibit lack of antigen-dependent tissue regeneration.

Microdomain (Lipid raft) Lipid raft is a cholesterol-enriched microdomain in the cell membrane, and is also called "caveola" in case it contains caveolin. It can be isolated from cells as low-density, detergent-insoluble membrane fractions, which are rich in cholesterol, sphingolipid, glycosphingolipid, GPI-anchor protein, and a variety of signaling molecules. It has been reported that it is involved in various cellular events such as signal transduction, apotosis, endocytosis, infection, and membrane trafficking.

Midkine Midkine, a heparin-binding growth factor or cytokine, promotes cell survival and cell migration, and is deeply involved in cancer progression, the onset of inflammatory diseases, and the preservation and repair of injured tissues. Midkine also is involved importantly in development and is strongly expressed during midgestation. In the adult, midkine expression is strongly induced during oncogenesis, inflammation, and repair. Midkine binds to the trisulfated unit of heparan sulfate and chondroitin sulfate E unit, and the binding is essential for expressing its activity.

Milk oligosaccharides Mammalian milk usually contains, in addition to lactose as a dominant saccharide, a variety of other saccharides, called milk oligosaccharides, which commonly have a lactose unit at their reducing end. For example, human milk contains more than 100 milk oligosaccharides. The oligosaccharide content of mature human milk is from 12 g to 14 g per liter. Milk oligosaccharides act as anti-microbial defense factors against pathogenic bacteria. They appear to firstly act as prebiotics, stimulating the growth of beneficial organisms such as bifidobacteria, and secondly act as receptor analogues, competing with pathogenic bacteria for attachment to gastrointestinal receptor sites.

Molecular dynamics calculation Molecular dynamics calculation is a computer simulation where atoms and molecules are allowed to move under Newton's laws, giving a view of the motion of the atoms. The motion of atoms (coordinates, velocity, and acceleration) is calculated by solving Newton's laws iteratively (every one femto (10–15) seconds). The term, MD (molecular dynamics) simulation is also used for molecular dynamics calculation.

Monoclonal antibody Monoclonal antibodies are identical immunoglobulin molecules that are produced by cloned antibody-producing cells. Most monoclonal antibodies are produced by hybridomas, hybrid of B cells and myeloma cells. A lot of monoclonal antibodies against specific oligosaccharide antigens are produced and used for the analyses of oligosaccharides.

Mucin-type glycans O-linked glycans are usually attached to Ser/Thr residues of the core protein through the GalNAc residue at the reducing end. These are usually referred to as "mucin-type" glycans because of their predominant occurrence on the mucus glycoproteins.

Multi-dimensional HPLC mapping method Multi-dimensional HPLC mapping is a method to determine the structures of N-linked oligosaccharides by HPLC. N-linked oligosaccharides are released from proteins, and the reducing ends of released oligosaccharides are fluorescently labeled with 2-aminopyridine. Pyridylaminated oligosacchardes are first separated with their charge using the DEAE column. Next, each fraction is separated sequentially by HPLC using ODS and amide columns. The structure of PA-oligosaccharide can be deduced by comparing its elution position with that of the reference PA-oligosaccharide on the map.

Multistep tandem mass spectrometry   Tandem mass spectrometry is performed by: (1) selection of a precursor ion, (2) collision of selected ions with rare gas, (3) separation of fragment ions, and (4) detection of fragment ions. Multistep tandem mass spectrometry is a method used to repeat the selection and collision of fragment ions. Detailed information on the structure of biomolecules (protein, oligosaccharides) can be obtained.

Myelin   Myelin is an electrically insulating dielectric phospholipid layer that surrounds the axons of many neurons. The insulating properties of the compact myelin around the intermodal segment of the axon enable the fast propagation of electrical signals at the myelin-free node of Ranvier. It is a product of Schwann cells in the peripheral nervous system and of oligodendrocytes in the central nervous system. The high ratio of lipids to proteins is unique for myelin and differs from the ratio seen in other membranes. Specifically in myelin, glycosphingolipids are enriched. The myelin glycosphingolipids consist of two major components, galactosylceramide and its sulfated form.

***N*-acetylglucosamine 6-*O*-sulfotransferase**   *N*-acetylglucosamine 6-*O*-sulfotransferase, often abbreviated as GlcNAc6ST, is a sulfotransferase, which transfers sulfate onto GlcNAc residue, forming 6-*O*-sulfated GlcNAc structure. There are various molecular species of GlcNAc6STs, with different tissue distribution and substrate specificity. This enzyme is involved in the biosynthesis of L-selectin ligand as well as keratan sulfate. A sulfate group of L-selectin ligand in peripheral lymph nodes are synthesized by the action of both GlcNAc6ST-1 and GlcNAc6ST-2.

Neoglycolipid   Neoglycolipid, a generic name of glycolipid with the structure being nonexistent or not found in nature, is synthesized by artificially replacing the sugar moiety of gangliosides or sphingoglycolipids. The one in which the structure of the lipid moiety was converted to the non-natural type is often included in this category.

Neoglycoprotein   Neoglycoprotein, a generic name of glycoprotein with the structure being nonexistent or not found in nature, is derived by chemically inducing sugars to proteins. Most neoglycoproteins are generated by covalently bonding sugars to the lysine amino groups on the side chain of the protein.

Neoproteoglycan (probe)   Neoproteoglycan, a kind of artificial glycoconjugates, is synthesized by covalently bonding polysaccharide chains such as glycosaminoglycans to the proteins. Neoproteoglycans are used for the search, purification, analyses of the interactions and functional substitutions, etc. of the substances with the affinity as the models of proteoglycans.

Neuroglycan C   Neuroglycan C, which is abbreviated to NGC, is a transmembrane chondroitin sulfate proteoglycan that is exclusively expressed in the central nervous system. The human NGC gene is written as CSPG5, and is located on chromosome 3p21.3. It is known that gene expression changes are elicited by drug addiction and nerve injury, NGC activates epidermal growth factor (EGF) receptors, ErbB2 or ErbB3, and regulates neurite outgrowth. Studies in knockout mice have suggested that NGC regulates synapse formation.

NKT cells   NKT cells are a subset of T cells, which express both TCR (T cell receptor) and NK cell receptor. They are involved in self-tolerance, allograft rejection, and cancer immunity. Galα1-containing glycolipids (Galα1-ceramide, Galα1-3Galβ1-4Glcβ1-ceramide) are presented by CD1d, an MHC class I-like molecule on antigen-presenting cells, and therefore cells recognize the complex with TCR, leading to their activation. Much attention has been paid to NKT cells with respect to their relationship of glycans with cancer immunity, and self-tolerance.

N-linked glycans *N*-linked glycans is one of the most common glycosylation types in eukaryotes, linked to the side chain of Asn residues in proteins. In eukaryotes, *N*-glycosylation only occurs at the consensus sequence called sequon, (Asn-X-Ser/Thr; X is any amino acids except Pro), whereas there is a rare occurrence of *N*-linked glycans at the Asn-X-Cys sequence.

Notch signal The Notch signaling pathway controls multiple cell differentiation processes during the developmental stages. Notch receptor interaction with a ligand (Delta-like, jagged in mammals) on neiboring cells induces proteolytic cleavage and release of the intracellular domain, which then moves to the nucleus to alter gene expression. The Notch extracellular domain has EGF-like repeats (e.g., Notch 1 has 36 of these repeats). Each EGF-like repeats can be multiply modified with *O*-fucose and *O*-glucose. *O*-Fucose can be elongated. Without *O*-fucosylation, Notch signal is not propery transduced.

Nucleotide sugar transporter Nucleotide sugar transporters are hydrophobic proteins in the Golgi apparatus or endoplasmic reticulum. They specifically antiport nucleotides sugars pooled in the cytosol into the lumen of Golgi or endoplasmic reticulum with the corresponding nucleoside monophosphates. With the molecular identities of nucleotide sugar transporters being unveiled by cloning in 1996, it has been reported that various nucleotide sugar transporters have different specificities in many eukaryotes. They present differently in different types of organisms; GDP-mannose transporter is present in plant, yeast, and protozoan, whereas it is absent in mammal.

O-Fuc glycans *O*-Fuc glycans is a type of *O*-glycans, where Fuc residues are attached to Ser/Thr residues in glycoproteins. There are two distinct types of *O*-Fuc modifications; one attached to EGF-like domain, and the other attached to thrombospondin type I repeat domain. The elongated structures of *O*-Fuc glycans are known to be different between these two types.

Oligosaccharide processing The term, oligosaccharide processing, implies the process of oligosaccharide biosynthesis. In general, oligosacchairdes are synthesized by a series of glycosyltransferases using sugar nucleotides as the donor substrates in the Golgi complex. Although the processing pathways of *O*-glycans, *N*-glycan, and proteoglycan are different, each glycan is in principle synthesized with specific glycosyltransferases.

Oligosaccharide profiling Oligosaccharide profiling is to characterize the oligosaccharide structures of the sample. The profiling is performed by identifying the oligosaccharides with liquid chromatography, electrophoresis, or lectins.

O-linked glycans See "*O*-linked glycopeptides".

O-linked glycopeptides Defined narrowly, *O*-linked glycopeptides stand for the ones containing mucin-type *O*-linked glycans (Ser/Thr-linked GalNAc). In a broad sense, they can include any glycopeptides containing *O*-linked glycans (not only *O*-GalNAc, but other type of *O*-linked glycans such as *O*-Fuc, *O*-Man, *O*-GlcNAc, *O*-Glc, etc.).

O-linked *N*-acetylglucosamine In this type of glycosylation, GlcNAc is linked to the hydroxyl group of Ser or Thr residues on proteins. *O*-linked *N*-acetylglucosamine is found in cytoplasm/nuclear proteins and an interplay between *O*-GlcNAc and phosphorylation has been suggested.

O-Man glycans *O*-Man glycans is a type of *O*-glycans, where Man residues are attached to Ser/Thr residues in glycoproteins. It is commonly observed in yeast, but similar modifications can be found in mammalian cells. The terminal structures of *O*-Man glycans are quite distinct between yeast and mammalian cells. This modification is essential for viability in yeast, whereas the deficiency of *O*-Man modifications results in congenital muscular dystrophy in humans.

One-pot glycosylation One-pot glycosylation method refers to one in which several glycosyl donors are allowed to react sequentially in the same flask, resulting in a single main oligosaccharide product. This method integrates several glycosylation steps into one synthetic operation to furnish target oligosaccharides in a short period of time without the need for protecting group manipulation and intermediate isolation. Therefore, this method is an attractive method in solution-phase methodology for the synthesis of oligosaccharide library.

Ozonolysis Ozonolysis is a chemical reaction, which selectively cleaves olefins under mild conditions. This method can be applied to release oligosaccharides from glycosphingolipids. After ozone gas is saturated into the glycoshingolipid solution, it is evaporated to dryness at room temperature. Alkaline treatment of the residue causes the elimination reaction to liberate intact oligosaccharides, whereas the ceramide portion is destroyed.

PAPS transporter It transports PAPS (3′-phosphoadenosine 5′-phosphosulfate) from the cytosol into the lumen of the Golgi apparatus. PAPS is synthesized by PAPS synthetase in the cytosol; therefore, the activity of PAPS transporter is essential for the sulfation of proteins/glycans in the Golgi. The gene was cloned from both humans and fruitfly in 2003.

Paroxysmal nocturnal hemoglobinuria Paroxysmal nocturnal hemoglobinuria is an acquired hemolytic anemia caused by somatic mutation of PIG-A gene in hematopoietic stem cells. PIG-A gene encodes a protein required for the synthesis of the GPI anchor. The gene that codes for PIG-A is inherited in an X-linked fashion, which means that only one active copy of the gene for PIG-A may exist. If a mutation occurs in the PIG-A gene in a bone marrow stem cell, it leads to a defect in the GPI anchor in the blood cells. Several GPI-anchored proteins in the blood cell, protecting the cell from destruction by the complement system, are deleted by PIG-A mutation. Deletion of these proteins causes a complement activation by some infection, which leads to the destruction of red blood cells.

Part-time proteoglycan Proteoglycans represent glycoproteins to which sulfated glycosaminoglycan is attached covalently. In case of part-time proteoglycan, some portions of core protein without having glycosaminoglycan chains also exist. Amyloide-beta precursor protein (APP), neuroglycan C, and thrombomodulin are typical part-time proteoglycans.

Pattern recognition In case of antibody-dependent pattern recognition, antigen epitope that consists of a couple of amino acid residues binds to antigen-binding site of the antibody (paratope) just like a lock-and-key model. Innate immune responses are initiated by pattern recognition receptors (PRRs), which recognize three-dimensional structures, consisting of a repeating domain of specific low molecular component, such as lipopolysaccharide (LPS) of Gram-negative bacteria.

Peptide mapping In this technique, a particular protein is digested by chemical method or proteolytic enzyme(s) of known specificity, and the peptide fragments are separated by HPLC or electrophoresis. This technique can provide useful information about the amino acid sequence and posttranslational modification of the protein of interest.

Phytosphingosine Sphingosine (2-amino-4-octadecene-1,3-diol) is an 18-carbon amino alcohol with an unsaturated hydrocarbon chain, which forms a primary part of sphingolipids, a class of cell membrane lipids that include sphingomyelin. Sphingosine derived from plants is called phytosphingosin, which has a C-4 hydroxyl group instead of unsaturated carbon and thereby forms a more stable complex structure with additional hydrogen bonds.

PNGase (peptide:N-glycanase) PNGase is also called glycoamidase or N-glycanase. The enzyme cleaves the amide bond between N-glycans and the linkage Asn residues (it should be noted that the enzyme is not a "glycosidase", but "amidase": as a result, the linkage Asn residues are converted to Asp residues because of deamidation reaction). This enzyme, of plant and bacterial origin, has been widely used for structure/function analysis of N-glycans on glycoproteins. Later, this enzyme was found to be ubiquitously observed in the cytosol of eukaryotic cells, and the cytosolic PNGase has been shown to be involved in the process of quality control system for newly synthesized proteins, called ERAD (ER-associated degradation).

Polylactosamine Polylactosamine is a linear polymer of N-acetyllactosamine, Galβ1,4GlcNAc. It is also known as an i antigen, one of the blood group antigens. Polylactosamine is found in diverse glycoproteins and glycolipids, and various functional glycans containing Lewis X antigens are added to its termini. Polylactosamine itself is involved as a ligand of Galectin.

Polypeptide-N-acetylgalactosamine transferase Polypeptide N-acetylgalactosamine transferase is involved in the catalysis of a single glycosidic linkage, GalNAc1-O-Ser/Thr, and collectively the GalNAc-transferase isoforms control the initiation of mucin-type O-glycosylation. It is also called pp-GalAcTase, pp-GalNAc-T, and so on. Polypeptide GalNAc transferase gene family contains 20 genes, of which 15 have been shown to encode functional enzymes. The GalNAc-transferase gene family is the largest mammalian family of glycosyltransferases.

Polysialic acid Polysialic acid (PSA) is a polymer of sialic acid whose degree of polymerization (DP) is 8–200 Sia residues. The most common structure of PSA is the Neu5Ac polymer, whose inter-residual linkage is α2,8. PSA occurs in NCAM, voltage-sensitive sodium channel, capsular polysaccharides of meningitis bacteria, trout polysialoglycoprotein (PSGP), and CD36. PSA has an anti-adhesive effect in cell–cell and cell–

extracellular matrix interactions because of its bulky volume and anionic nature and thereby is involved in the neural cell migration.

Post-translational modification Post-translational modification is the chemical modification of a protein after its translation. Examples are phosphorylation, glycosylation, glycation, and protease cleavage. In general, most of the proteins are synthesized into mature ones by attaching to the functional groups.

Prediction of oligosaccharide modification and interaction Modification of *N*-linked oligosaccharides can be predicted by the primary and tertiary structures of the proteins. The mode of interaction between oligosaccharide and protein can also be predicted with a computational method using the primary and tertiary structures of the sugar-binding proteins.

Proteasome Proteasomes are gigantic protein complexes inside occurring eukaryotes and archaes, as well as some bacteria. In eukaryotes, they are located in the nucleus and the cytosol. The most common form of the proteasome is known as the 26S proteasome, which is about 2000 kilodaltons (kDa) in molecular mass and contains one cylinder-shaped 20S core particle, which exhibits various protease activities, and two 19S regulatory subunits (PA700). There is also another regulatory subunit, called 11S regulatory subunit (PA28), and combinations of different regulatory subunits generate various distinct proteasomes. They are involved in not only proteasomal degradation of unwanted proteins, but also in other biological processes such as antigen processing.

Protein *O*-mannose β-1,2-*N*-acetylglucosaminyltransferase Protein *O*-mannose β-1,2-*N*-acetylglucosaminyltransferase (POMGnT1) catalyzes the formation of GlcNAcβ1-2Man linkage of *O*-mannosyl glycans. The *POMGnT1* gene is a causative gene of muscle-eye-brain disease (MEB), which is one of the congenital muscular dystrophies.

Protein *O*-mannose transferase Protein *O*-mannose transferase catalyzes the transfer of mannose to serine/threonine residues of proteins using dolichol phosphate-mannose (Dol-P-Man) as a donor substrate. This enzyme activity requires two homologs, POMT1 and POMT2. The *POMT1* and *POMT2* genes are the causative genes of Walker-Warburg syndrome (WWS) categorized into congenital muscular dystrophies. Knockout mouse lacking *POMT1* is embryonic lethal.

Proteoglycans Proteoglycans represent a special class of glycoproteins in which more than one glycosaminoglycan (GAG) chain is covalently attached to the core protein. Proteoglycans can be categorized into heparan sulfate-, chondroitin sulfate-, dermatan sulfate-, and keratan sulfate proteoglycans depending on the nature of GAG chains. Proteoglycans are expressed on the cell surface or represent the extracellular matrix components.

P system blood group antigen It is a blood group antigen determined mainly glycan core structures of globo-type glycolipids. Whereas P antigen, globotetrasylceramide (Gb4), is accumulated in P+ blood group, its precursor globotriaosylceramide (Gb3), is accumulated in P- (P-minus) blood group, and is referred to as the pk type. Deficiency in Gb3 synthesis is called little p (p), where P-antigen is absent. Occurrence of pk and p type is rare. On the other hand, there is another type of P-antigen, denoted as P1/P2, and 80% of Caucasians contains PI antigens. The antigenic structure of P1 has been assumed to be paragloboside (neolactotetraosyl-ceramide), where terminal Galα1-4 structure is observed. It has been found that Gb3 synthetase is also involved in the bio-synthesis of P1 antigen, and in P2 the level of expression of this enzyme is found to be at low level.

PTPζ PTPζ is one of the receptor-type protein tyrosine phosphatases of the 21 so far found. Also known as RPTPβ, it is synthesized as a chondroitin sulfate proteoglycan. A protein formed from the extracellular domain of PTPζ through selective splicing is called phosphacan. PTPζ binds to its ligands such as pleiotrophin, a heparin-binding growth factor, or midkine, and mediates diverse signal transduction events.

Pyridylamination (PA sugar) In this pyridylamination method, reducing termini of free oligosaccharide chains are modified with 2-aminopyridin. This modification has advan-tages of not only enhancement of the sensitivity of detection by HPLC and mass spec-trometry analysis, but also better separation of PA-labeled oligosaccharides in the reversed phase HPLC.

Reducing end of sugar chain The reducing end of a sugar chain is the name given to the direction of a sugar chain sequence. One side with an anomeric carbon is called the reducing end; on the other hand, the opposite side is called the non-reducing end. There is more than one non-reducing end in the case of branching sugar chains.

Regenerative medicine Regenerative medicine aims to recover the function of organs or tissues, which cannot be regenerated spontaneously. There are several methods to regenerate organs and tissues: (1) Regenerate the isolated tissues ex vivo and re-inject afterwards. (2) Extract the self-regenerating activities in vivo by providing growth hormone, inducing genes, or giving scaffolds for regeneration. (3) Differentiate the embryonic cells or tissue stem cells to intended organs or tissues.

RNAi RNA interference: the introduction of double-strand RNA (dsRNA) into cells causes its degradation by RNase III-like enzyme, dicer, into 21–25 bases short interfering RNA (siRNA). The siRNA dissociates to single-strand RNA by the action of RNA heli-case, and the single-strand RNA is incorporated into the protein complex called RICS (RNA-induced silencing complex). RISC then recognizes the homologous mRNA with the single-strand RNA, degrades them, leading to gene silencing at the mRNA level.

Saccharide primer A saccharide primer is composed of an alkyl chain and a carbo-hydrate part. Saccharide primers in the culture medium are incorporated into the cells. The primer is a target of glycosyltransferases and the sugar chain is extended on the primer, whose structure is dependent on the biosynthetic pathway of the cell used in the study. Most saccharide primers coordinated with sugar chains are secreted into the medium. The structure of the carbohydrates attached on the primer is similar to that of the oligosaccharides in the cell. It is possible to extend the sugar chain on the primer in a cell-free system.

Scavenger receptor Scavenger receptor is mainly found in macrophages as a receptor of oxidized LDL. Scavenger receptor recognizes negative charges properly arranged on molecular surface, and binds to various molecules including mucins.

Selectin Selectin is a family of cell adhesion molecules that mainly recognize carbohydrates as a ligand. The family consists of three members, E-, P- and L-selectin. E-selectin is expressed in vascular endothelial cells while P-selectin is expressed in vascular endothelial cells and platelets. E- and P-selectins mediate the adhesion between these cells and ligand-positive cells such as leukocytes. L-selectin is expressed on leukocytes and mediates the interaction between leukocytes and ligand-positive cells.

Seminolipid Seminolipid is a sulfoglycolipid expressed specifically in permatocytes and subsequent germ cells. The carbohydrate moiety of seminolipid is galactose-3-sulfate, and the acyl moiety is alkylacylglycerol. The carbohydrate moiety of seminolipid is the same as that of sulfatide and synthesized by common enzymes, ceramide galactosyltransferase and cerebroside sulfotransferase. More than 90% of glycolipids in the testis consists of seminolipid.

Sialic acid Sialic acid is a generic term for *N*-acetylneuraminic acid and its derivatives, a nine-carbon monosaccharide. The variations of the derivatives are made by the acetylation of hydroxyl group, deletion of *N*-acetyl group, conversion of *N*-acetyl group to *N*-glycolyl group, and so on. It shows acidity because of its carboxyl group. Sialic acid is attached to non-reducing end of the sugar chain, by the action of sialyltransferase using CMP-sialic acid as a donor. Sialic acid is synthesized from *N*-acetylglucosamine through *N*-acetylmannosamine.

Sialidase Sialidase is an exo-type α-glycosidase that releases sialic acid from nonreducing terminals of glycans of glycoproteins, glycolipids, and oligosaccharides. It is also called neuraminidase. Sialidase was first discovered in a virus as a receptordestroying enzyme, and has been found in various tissues of microorganisms (bacteria, protozoa) and vertebrates. Four kinds of animal sialidases are identified with different cellular localizations and zymological properties, and they are known to control a lot of cellular functions.

Sialyl 6-sulfo Lewis X Sialyl 6-sulfo Lewis X is a sugar chain containing NeuAcα2→3Galβ 1→4[Fucα1→3]GlcNAc(6-sulfate)β1→R). Sialyl 6-sulfo Lewis X was first identified as a ligand for L-selectin. The binding affinity of L-selectin is strongly dependent on 6-sulfo group. It also acts as a ligand for E- and P-selectin.

Sialylated glycan Sialylated glycan is a sialic acid-containing glycan and is distributed especially in proteins and gangliosides. Sialylated glycan on the cellular surface is a target for various pathogens including influenza virus. On the other side, it is also a ligand for animal lectins. Several sialylated glycan-binding lectins were reported, such as selectins and siglec proteins. Secletins are involved in the accumulation of leukocytes at the inflammation site. Siglec proteins are known to control immune responses.

Sialyltransferase Sialyltransferase transfers a sialic acid to glycans attached onto proteins and lipids using CMP-NeuAc (or CMP-NeuGc) as a donor. A total of 20 types of sialyltransferases are identified in human and mouse. Sialyltransferases are roughly classified into four groups: ST3Gal-, ST6Gal-, ST8Sia-, and ST6GalNAc-. All sialyltransferases have a common sialyl motif: L, S, and VS. The motif is highly conserved, and is successfully utilized for the PCR cloning of sialyltransferases.

Siglec family Siglec family binds to sialyl oligosaccharides and is a member of the immunoglobulin superfamily. Siglec family is a type I membrane protein. Siglec binds to sialylated glycans using its extracelluer domain, and the signal transduction is mediated by the intracellular domain. Carbohydrate recognition and signal transduction are correlated with each other, but the regulatory mechanism is not clarified completely. The knockout mouse lacking siglec 1–4 is established.

Single molecule fluorescent measurement (fluorescent correlation spectroscopy) There are various methods of fluorescent measurement at a single molecule level by confocal or two photon microscopy, such as fluorescent correlation spectroscopy (FCS), fluorescence intensity distribution analysis (FIDA), or fluorescence polarization (FP) measurement. FCS measures fluorescent intensity fluctuation (translational diffusion time) in an extremely small volume of sample, allowing the determination of the molecular size of the fluorescent molecules. Using this method, real-time, high-speed and high-precision analysis can be achieved for measurement of various molecular parameters, including size, concentration, fluorescence brightness, and the respective amounts of different molecules present, thereby assessing molecular interactions between fluorescent molecules and other molecules in solutions. See also "Fluorescence polarization measurement".

Site-specific glycan structure Glycoproteins having multiple glycosylation sites may contain site-specific glycan structures. Such a difference is because of the fact that processing of glycan chains are affected by protein conformation. Determination of site-specific glycan structure therefore provides us useful information regarding protein structure–function relationship.

Solid-phase carbohydrate synthesis An automatic solid-phase synthesizer for peptide or DNA oligomer is available. In the same way, solid-phase carbohydrate synthesis attempts speeding up by performing the synthesis of sugar chain on a solid-phase support. The requirement of a variety of blocked monosaccharide units and the regulation of stereochemistry of glycosylation are still problems, because all of the oligosaccharides are not linear and thus different from the peptide of DNA. Using enzymatic reaction is also effective. The affinity of solid-phase support for inorganic solvent and the availability of enzyme and nucleotide sugar are keys to the synthesis.

Spermatogenesis Spermatogenesis occurs in the seminiferous tubules in the testis. Spermatogonia, a kind of germline stem cells, are found in the seminiferous tubules. Spermatogonial stem cells differentiate into spermatocytes, and subsequently undergo meiosis twice, differentiating into haploid cells referred to as spermatids, and migrating from the surrounding area to the lumen of the seminiferous tubules on the stroma cells, termed Sertoli cells, with mutual interaction. After that, they mature into spermatozoa via morphogenesis. A series of these steps is called spermatogenesis.

Sphingolipid ceramide *N*-deacylase Sphingolipid ceramide *N*-deacylase (SCDase) hydrolyzes the *N*-acyl linkage between fatty acids and sphingosine bases in the ceramide moiety of various glycosphingolipids and sphingomyelin. It hardly acts on the free ceramide, and the primary structure of SCDase is quite different from the ceramidase that acts only on free ceramide. The enzyme also catalyzes the condensation reaction (reverse hydrolysis) between fatty acids and sphingosine bases to form sphingolipids.

Stable isotope labeling Labeling with stable isotopes (^{2}H, ^{13}C, ^{15}N, etc) is an important method for the conformational analysis of biomolecules by NMR. Stable isotope labeling is also applied to the structural analysis of oligosaccharides by mass spectrometry. It is performed by: (1) addition of labeled metabolic precursor into the culture medium, (2) chemical synthesis, and (3) in vitro enzymatic reaction.

Stem cell In general, stem cells are defined as cells, which retain the ability to: (1) grow; (2) renew themselves through mitotic division; (3) differentiate into a diverse range of specialized cell types; and (4) repair damage of specific tissues/organs. The two major types of stem cells are: embryonic stem cells (ES cells) that are derived from blastocysts, and somatic stem cells found in adult tissues. Some of the example of somatic stem cells are: hematopoetic stem cells; neuronal stem cells, and hepatic stem cells. They are used for the maintenance and/or repair of respective cells or tissues.

Structural biology Structural biology aims to determine the three-dimensional structures of biological macromolecules at atomic resolution by using X-ray crystallography and nuclear magnetic resonance (NMR) spectrometry and to elucidate the molecular mechanism of functional expression from the structural point of view.

Sugar chain remodeling Sugar chain remodeling is used to change the structure of a sugar chain by artificial manipulation, such as overexpression or gene knockout, of genes of glycosyltransferases. A biological function of sugar chain could be to change dramatically or not after this manipulation.

Sugar chip Sugar chip is prepared by immobilizing the oligosaccharides onto the metal-coated glass plate (typical metal is gold). Sugar chip is used for the surface plasmon resonance (SPR) analysis of the interaction between oligosaccharides and oligosaccharide-binding proteins.

Sugar library Sugar library is a set of oligosaccharides and their derivatives obtained by organic synthesis, enzymatic synthesis, or isolation from natural resources. Sugar library is a powerful tool in the glycomics area. The library can be used as standard samples of structural analysis and also applied to the analysis of sugar–protein interactions combined with the microarray techniques.

Sugar oxazoline Sugar oxazoline is produced by the dehydration and condensation of the acetamido group (C2) with hydroxyl group (C1). Sugar oxazoline is widely used for the chemical and enzymatic glycosylation of 2-acetamido-2-deoxy sugars.

Sugar–protein interaction With structural biology having developed, the mechanism of interaction between proteins, such as lectin and antibody, and sugars has been visualized at the atomic level. Sugar–protein interaction is mainly because of hydrogen bond and van der Waals interactions, etc., which is generally weaker than protein–protein interaction. Sugar–protein interaction has often achieved high affinity and specificity through multivalent binding.

Sulfated glycan Sulfated glycans are any of the groups of glycan-attached sulfate group. Addition of the sulfate group is synthesized with various sulfotransferases. Sulfated glycans have been found in the backbone glycan such as N-linked or O-linked glycan, glycosaminoglycan, and glycolipids. They are involved in significant biological processes such as development and immunity.

Sulfotransferase Sulfotransferase is a transferase enzyme, which acts on a sulfate group. These enzymes utilize activated sulfate, PAPS (3′-phosphoadenosine-5′-phosphosulfate), as a high-energy donor, and act to transfer a sulfate group to a second substance such as protein, glycan, steroid, and tyrosine. These enzymes can be classified into two groups, those localized in the Golgi membrane and those in the cytosol. The enzyme related to sulfonation of carbohydrate is localized in the Golgi membrane.

Surface plasmon resonance In order to excite surface plasmons in a resonant manner, one can irradiate an electron or light beam to the planar metal interface in an appropriate condition, and surface plasmon waves that propagate parallel along a metal interface are resonant with the evanescent wave. Because evanescent waves are affected by the refractive index of the material onto the metal surface, this phenomenon is used in many biosensor applications. For instance, one can measure receptor–ligand interaction by adsorbing the ligand with the metal surface.

Syndecan Syndecans are a family of membrane-bound heparan sulfate proteoglycans. Structurally similar syndecans, syndecan-1 to -4, were reported. Syndecans are involved in a variety of functions, acting as co-receptors of growth factors, integrins, and fibronectins.

T-antigen T-antigen is a cancer-related glycan antigen, Galβ1-3GalNAc-Ser/Thr. It is also called TF antigen, after Thomsen and Friedenreich who discovered this antigen.

Tertiary structure model The molecular structure is described with constituent atomic positions using three-dimensional coordinate system (x, y, and z). Although NMR spectroscopy and X-ray crystallography do not directly provide the atomic position of the molecule, molecular structures determined by them are called "tertiary structure model".

Therapeutic antibody Antibody is produced when exogenous antigen (for example, pathogenic bacteria) invades into the body. Antibody is a component of the immune system that prevents pathogens from entering or damaging the cells. Antibody therapeutics is aimed to utilize the functions of antibody in the immune system. Therapeutic antibodies against TNFα, EGF and Her2 are now used for intractable inflammatory disease and cancer. It has recently been shown that removal of core fucose on human IgG1 oligosaccharide results in 50–100-fold enhancement of antibody-dependent cellular cytotoxicity (ADCC). Hence, control of the IgG fucosylation could be one of the most promising technologies to improve the efficacy of therapeutic antibodies.

Thioester method Conventional solid phase peptide synthesis is limited by the peptide length. In order to prepare long polypeptides, it is necessary to couple short peptides. Thioester method is a technique used for creating long peptides and proteins: (1) a peptide containing thioester bond at its C-terminus is prepared. (2) The thioester is selectively activated by argentite. (3) The activated peptide is coupled to the N-terminal amino group of the other peptide. It is not necessary to protect the functional groups except for the amino groups (Lys, Arg) and thiol group (Cys).

Transferrin Transferrin is an iron-binding glycoprotein composed of 679 residues and possesses two Fe^{3+}-binding sites. Human serum contains 2–3 mg/ml of transferrin. Transferrin expresses biantennary oligosaccharides with N-acetylneuraminic acid at Asn413 and Asn611. Molecular weight of transferrin with oligosaccharides is 79,593. Transferrin is used as a marker molecule to examine the oligosaccharide variants involved in various diseases, such as CDG or chronic rheumatoid arthritis.

Transglycosylation Glycosidase has an ability to release the sugar by hydrolyzing the glycosyl-bond of substrates. The activity is also interpreted as transfer of sugar to water. *Transglycosylation* is a reaction in which sugar is transferred to the compound having hydroxyl group. The catalytic activity of the enzyme is dependent on the tertiary structure of the active site. A variety of oligosaccharides are now synthesized by transglycosylation reaction of exoglycosidases.

Trypanosoma *Trypanosoma* is a group of parasitic protozans. Sleeping sickness and Chagas's disease are caused by *Trypanosoma*. In sleeping sickness, the surface of blood cells is covered with GPI-anchored variant surface glycoprotein. The isoform of variant surface glycoprotein can be switched, dependent on the production of antibodies, resulting in the conversion of antigenicity. The structure and biosynthesis of GPI was first studied by using *Trypanosoma* because the GPI biosynthesis of *Trypanosoma* is highly active.

Tumor marker Tumor markers are produced in tumor cells and can be detected in body fluids such as serum. Examination of tumor marker is useful not only for the diagnosis of cancer and estimation of the primary focus, but also for monitoring therapeutic efficacy and postoperative follow-up.

Ubiquitin ligase A uniquitin ligase, also denoted as E3 enzyme, operates in conjunction with an E1 (ubiquitin-activating enzyme) and an E2 (ubiquitin-conjugating enzyme) to attach ubiquitin to an ε-amino group of lysine side chain(s) on a target protein. The system called "ubiquitin system" comprises E1–E3, and is involved in not only protein degradation, but also in quite diverse biological phenomena. A variety of E3 enzymes with different specificity is known to be responsible for the diverse recognition of substrates by them (e.g., a type of E3 enzyme, Fbs1/2, recognizes *N*-glycans).

Ultra-high field NMR At present, ultra-high field NMR spectrometers beyond 900 MHz are available owing to the development of new superconducting materials. Ultra-high field NMR spectra with high resolution and sensitivity offer an opportunity for detailed structural analyses of oligosaccharides and glycoproteins.

Verotoxicin/Shiga-like toxin Verotoxin, AB5-type toxin having the A subunit and five B subunits, is produced by the pathogenic *Escherichia coli* (*E. coli*) such as *E. coli* O157:H7. It is named after its ability to kill Vero cells. Shortly after, the verotoxin was referred to as Shiga-like toxin because of its similarities to Shiga toxin. Verotixin specifically binds to globotriaosylceramide (Gb3/CD77).

ZP3 ZP3 is a major glycoprotein component of the zona pellucida, a thick layer surrounding the mammalian egg. A glycan part of ZP3 plays an important role in binding to the sperm and therefore ZP3 serves as a sperm receptor on fertilization.

Index